Technical Writer's Freelancing Guide

PETER KENT

Sterling Publishing Co., Inc. New York

Library of Congress Cataloging-in-Publication Data

Kent, Peter.
    Technical writer's freelancing guide / Peter Kent.
        p.   cm.
    Includes bibliographical references (p.   ) and index.
    ISBN 0-8069-5836-7
      1. Technical writing.   2. Freelance journalism.   I. Title.
T11.K44   1992
808'.0666–dc20   91-39288
                                                                       CIP

10 9 8 7 6 5 4 3 2 1

Published in 1992 by Sterling Publishing Company, Inc.
387 Park Avenue South, New York, N.Y. 10016
© 1992 by Peter Kent
Distributed in Canada by Sterling Publishing
% Canadian Manda Group, P.O. Box 920, Station U
Toronto, Ontario, Canada M8Z 5P9
Distributed in Great Britain and Europe by Cassell PLC
Villiers House, 41/47 Strand, London WC2N 5JE, England
Distributed in Australia by Capricorn Link Ltd.
P.O. Box 665, Lane Cove, NSW 2066
Manufactured in the United States of America
All rights reserved

Sterling ISBN 0-8069-5836-7

Contents

Introduction   9
     The Three-Step Method   9
     Three-Step vs. Go-It-Alone   11

I. Getting into Technical Writing   13

1. Technical Writers: "Just Passing Through"?   13
     Technical Writers Explain Things   14
     How Much Do Freelancers Make?   14, 15
     But Do They Get Respect?   17
     The Trouble with Technical Writers   17
     Good Writers and Good Documentation   19

2. How to Get Started in Technical Writing   20
     Diversity Means Someone Will Hire You   20
     What Sort of People Are Technical Writers?   21
     How People Get into Technical Writing   21
     The STC Can Help You   26
     Stressing Your Skills in Your Résumé   27

3. Teach Yourself Technical Writing   28
     College Courses   28
     On-the-Job Training   28
     Books   29
     The Society for Technical Communication   29
     Pointers to Producing Useful Documents   29

II. An Introduction to Freelancing   33

4. What Is Freelancing?   33
     Contractors/Freelancers, Job-Shoppers,
        Temporary Employees, Consultants   33
     Technical Service Agencies/Job-Shops   34
     Several Ways to Freelance   34
     Which Way Is Most Profitable?   36
     Why Do Companies Use Contractors?   36
     How Much Experience Do You Need?   36
     The Three-Step Method of Freelancing   37
     The Money vs. Job Satisfaction   38

5. The Advantages of Freelancing   39
     Freelancing Provides Variety   39
     Getting Paid Now for the Hours You Work   40

Easier to Find New Work	41
Freelancers Can Travel	41
Freelancing as a Stepping-Stone	41
More Job Security, Less Office Politics	42

6. The Disadvantages of Freelancing — 44

You Don't Get Any Benefits	44
You Must Have More Savings	44
No Long-Term Work Relationships	44
No One Points Your Way	45
No Established Pension Plan, No Vacations	45, 46
No One Will Train You	46
No Unemployment Pay, Maybe No Workers Compensation	47

7. What Makes a Good Freelancer? — 48

The Ability to Handle Money	48
The Ability to Handle Uncertainty	48
The Ability to Sell Yourself	49
"Gossip" Can Move You Ahead	49
Be Good at What You Do	50
The Ability to Handle Change, Learn Quickly	51
The Ability to Get On Well with People	51

8. How Much Do You Earn? — 52

How Much Do You Really Earn?	52
When and Why Calculate Your Hourly Rate?	53
Medical/Dental/Vision Insurance	54
Long-Term Disability Insurance	54
Term Life Insurance	54
Tax-Free Savings Plans	55
Employer Contributions	55
Benefits You Might Not Use	55
FICA	56
Work Sheet: How Much Must You Earn?	57

III. The First Step in Freelancing 59

9. Finding the Technical Service Agencies — 59

Contacting More Than One Agency	59
Some Agencies Have Specialties	60
The More Offers, the More Choice	60
Making the Agencies Compete for You	61
Frequent Job Offers	61
Avoiding Bad Contracts	61
Pointers for Finding Agencies	62
Build Your Own List	64
Preparing Your Mailing	64
What Goes on Your Résumé?	64
Build a Relationship with Agencies	65

Contents

10. Negotiating with the Agencies — 67
- How Agencies Operate — 67
- Questions to Ask the Agencies — 67
- Discussing Your Pay — 68
- What About Overtime? — 69
- Agency Medical and Disability Policies — 70
- Vacations and Sick Leave — 70
- Working Out of Town: Questions to Ask — 71
- A Job Lead Is Not a Job — 72
- What Is a Fair Rate? — 73

11. Unethical Agencies — 75
- Agencies Can Hurt You and Your Client — 75
- Eleven Reasons Why Companies Should Avoid Agencies — 75
- Some Agencies Don't Pay Good Rates — 77
- Agencies Can Prevent Your Being Rehired — 78
- How Companies Can Avoid Agency Problems — 79

12. The Interview — 80
- Information You Need Before the Interview — 80
- Asking Questions of Your Own — 80
- The Client Can Join Your Network — 81

13. Contracts — 83
- Sample Contracts — 83, 85, 86
- Types of Contracts — 87
- Twelve Considerations When Writing a Contract — 87

14. Buying Your Benefits — 89
- Replacing Employer-Provided Benefits — 89
- Medical Policies — 89
- Organization Health Insurance — 91
- Cutting Medical Insurance Costs — 92
- Disability Insurance — 93
- Term Life Insurance — 94
- Tax-Free Savings Plans — 94
- Credit Unions — 96

15. At Work — 97
- Being "Good Enough" to Freelance — 97
- Freelancers Are Expendable — 98
- How to Act at the Company Office — 98

16. The Great Overtime Debate — 101
- Hourly Employees and Overtime Pay — 101
- Salaried Employee Exemptions — 101
- When a Salary Is Not a Salary — 102
- Salaried Pay Between Contracts — 103

17. Preparing for Step Two — 105
- Upscaling Your Pay — 105
- Finding Independent Contracts — 105
- Begin Saving Money — 106
- Educate Yourself — 106

IV. The Second Step in Freelancing 107

18. Networking — 107
- Networking—"Using People"? — 107
- What Do You Need a Network for? — 107
- How to Build a Network — 108
- Job Interviewing When You Don't Need Work — 109
- Asking Others Where, Who, How Much — 109
- Helping the Agencies — 110

19. Looking for Work — 112
- Steps in Finding Work — 112
- Cold Calling — 113
- Verbal Contracts Aren't Worth Much — 115

20. Taxes for the Freelancer — 116
- Doing Your Taxes Yourself — 116
- When Is Mileage Deductible? — 117
- Deductible Commuting — 117
- Per Diems — 118
- Your Tax-Home — 118
- Does Your Home-Office Fit IRS Criteria? — 120
- Deductible Expenses — 120
- Moving Expenses — 122
- Independent Freelancer/Sole Proprietor Taxes — 123
- Schedule C Deductions — 124
- Tax Forms You May Need — 125
- Tax Publications That Answer Your Questions — 125
- To Incorporate or Not? — 126

21. Are You Really an Independent Contractor? What the IRS Has to Say — 127
- IRS Criteria for Independent Contractors — 127
- New Rules and Rumors — 129
- Become a Consultant — 130
- Start Your Own Agency — 130

V. The Third Step in Freelancing 133

22. Where Next? — 133
- Consulting vs. Contracting — 133
- A Set Fee: Charging by the Project — 133
- Estimating Your Fee — 134
- Getting Work as a Consultant — 135
- Using Direct Mail — 135

Appendixes 139

A. Contractors' Publications — 139
- Addresses and Descriptions of Publications. — 139, 141, 142
- Additional Benefits Memberships Offer — 139

B. Technical Service Firms — 144
- Addresses and Phone Numbers of Agencies — 144

C. Associations — 147
Addresses, Phone Numbers, and Descriptions of Associations for Technical Writers — 147

D. Contractor's Checklist — 149
Preparing for the Job Search (Checklist) — 149
Negotiating with Agencies (Checklist) — 149
On the Road (Checklist) — 150
Looking for Independent Contracts (Checklist) — 151

E. Bibliography — 152
Books About the Business — 152

F. Co-ops, Correspondence Courses, Courts, Insurance, and the IRS — 155
Addresses and Phone Numbers to Get Help — 155

Index — 157

Introduction

To get thine ends, lay bashfulness aside;
Who fears to ask, doth teach to be denied.
—ROBERT HERRICK (1591–1674)

Why are you reading this book? Maybe you are already a technical writer and want to find out how you can leave your job and work freelance. Perhaps you have worked with other freelancers or contractors, have heard about the money they make, envied their freedom and detachment from office politics. Maybe you are *already* a freelancer (or contractor, or job-shopper, or consultant) and are just looking for ways to improve business. Or maybe you are a journalist who is tired of the low wages and long hours, or a copywriter who needs a change of pace. Or perhaps you are just someone who wants to be a *writer* and is looking for a way to make a living. Whoever you are, I'll tell you how to get where you want to go.

The main topic of this book is how to use your technical writing skills to build a freelance career, but I will also explain how to get those skills in the first place. You will find out how much money technical writers make, what they do, and how they came to be technical writers in the first place (it's remarkably easy for a determined newcomer to enter the profession). If you are not already a technical writer, you will learn how to become one. If you *are* a technical writer, you will learn how to double or triple your income.

This is not a "quit work/risk everything/work freelance/make loads of money" book, however. I *will* tell you how to become a freelancer, and I *will* tell you how to make a lot of money. But I'm *not* going to tell you to risk everything, because you don't need to.

Systems described in most freelancing books have two major problems. The first is that you need a lot of money to get started. And if you don't have the money? Wait until you have saved enough, says the author of one of these books, or get a bank loan. How long will it take to save enough money to survive without an income for, say, five or six months? And how many people want to go into hock on a gamble?

The second problem is the risk involved. Leaving a steady job and using your own marketing skills to find contracts is dangerous. If you are not successful, you lose not only the money you saved (or borrowed) but also the money you would have earned had you been fully employed.

I'm going to explain why you don't have to start with large sums of money, and why you don't need to take any risks. In fact, your job security will actually increase—real job security, that is, not the illusion of security that so many of us have.

I will describe a Three-Step Method that makes it safe and easy to become a freelance writer. In the first step you will use the technical service agencies to find work for you. You can go straight to a high-paying job, without any time out of work. Follow my advice on how to deal with the agencies and you can increase your income 50, 60, or 70 percent, perhaps more. (My first agency contract increased my income by only 7 percent. But I learned how to deal with the "job shops" and my next contract increased my income 75 percent.)

Many people don't even know technical service agencies exist, or if they do they think of them as shady, fly-by-night operations. In fact most agencies are well-established and well-respected businesses. Some agencies have branches all over the country, even the world, and gross tens or hundreds of millions of dollars a year. Many can provide medical insurance (usually at a higher cost than most corporations charge their employees, but not always). And they generally pay every two weeks, some even every week. If you don't know how to deal with the agencies, you probably will still earn a reasonable sum, but if you know the ground rules, you can significantly increase your income.

Step One helps you save money and build the network you need for Step Two. In Step Two you move out on your own, make your own contracts and cut out the middleman. You are still a contractor or freelancer, but you find your own contracts, make your own deals, and reap the benefits. But if you never make it to Step Two, that's okay as well—many people, content with the gains made in Step One, never make it that far.

Those who reach Step Three have a different form of relationship with their clients, more of a consulting relationship. No longer just "a warm body," the consultant gets work because the client is looking for skill and experience. The consultant's clients want someone to trust, someone with a reputation for being the best in the business. Some writers in this third group make in the $100,000s.

But you don't have to be an expert to begin my Three-Step Method. Many entry-level people go straight into contracting; I know one writer, for example, who went straight from college to a position paying over $20 an hour. Unlike consultants, who are usually selling advice based on experience, contractors—those in Step One—are often filling spaces in an employer's work force, the proverbial warm bodies. Sometimes that space is waiting for an entry-level person.

What does a seventeenth-century poem about seduction have to do with freelancing? As you will discover when you read on, the key to success in freelancing is asking the right questions. You must ask people for information, for help, and for advice. And you must learn to ask for contracts, and ask for a high rate. Until you learn to ask for a lot of money for your services, you will not make the large sums that some earn. The contractors people think of as "pushy" are the ones making the most money. The meek may inherit the earth one day, but for the moment the confident—those who don't fear to ask—are the ones making the money.

Many books on freelancing try to cover everything, including the details of setting up a business—accounting, getting a business name, setting up your office, and so on. One book even tells you in which desk drawer you should place your most-used items (the top right one, apparently)! My book, however, concentrates on finding work. Where I have digressed into taxes and the law, I have done so to discuss issues of particular concern to freelancers. If you want a general business book, look elsewhere—there are many excellent ones in your local bookstore.

Some freelancing books are also too general in other ways; they purport to help *all* freelancers: accountants, artists, programmers, personnel managers, dancers, and so on. These professions all have different markets. I wrote this book for technical writers, though its principles apply to many technical professionals. If you are a computer programmer, an electronics engineer, an aircraft tool designer, or one of several hundred other technical professions that are placed by the technical service agencies, you can also use the methods described in this book. Some other professions, such as nursing, accounting, and business management, also have similar markets and have their own agencies that will sell their services, so many of the principles in this book apply to those professions; but I would never suggest that this book is "everyone's guide to freelancing."

You may not need to use everything in this book. I've used most, but not all, of the techniques in this book, to varying degrees. Some I wish I had used more, and some I would use more if I really had to—many writers will find work so easy to come by that they just don't need to dig around too much. For instance, if a day or two on the phone to friends is enough to find a job, you don't need to spend long hours cold-calling businesses. Some techniques I regret not using: I know my failure to send a change of address letter to all my contacts lost me at least one well-paid contract, for example.

Finally, let me show you a direct comparison between using my Three-Step Method and a typical "go-it-alone" freelance plan:

Three-Step Method	**Go-It-Alone Plan**
Increase your income immediately	Probably lose income at start
Partial success means an increase in income	Partial success means a loss of income
Start with minimal savings	Need substantial savings
Start with minimal expenses	Need a large investment
Start saving money in Step One	Lose money in first stages
Build a network in Step One	Must have a network before starting
Return to Step One when things get tough	Return to full-time employment if you fail
Psychologically easier way to start	On your own from day one

If you want to freelance, and don't know where to start, you've come to the right place. Read on!

I.
Getting into Technical Writing

1
Technical Writers: "Just Passing Through"?

There is a common theme in technical-writing humor: Technical writing is boring, technical writers get no respect, and everyone would rather be doing something else. A friend of mine publishes a small newsletter called *Dull Way (A Publication for Tech Writers & Dullards)*. Another friend claims he was dancing with a girl one night when she asked what he did for a living; when he said he was a technical writer she walked away. And the following poem by Roger L. Deen satirizes a common feeling among technical writers.

Just Passin' Thru

He claimed to be one of those few
Who detested what tech writers do,
So as the years rolled by,
He continued to cry
"I'll be gone soon—I'm just passin' thru."

Quoth he to all who would hear:
"Tech writin's for chumps, that's clear.
I'm smarter than you,
'Cause I'm just passin' thru."
He's been saying that now seven year.

When Judgment Day finally comes true,
And Saint Peter asks, "What'd you do?"
He'll answer with pride,
As he swaggers inside,
"I'm a tech writer, and I'm just passin' thru."

There's no smoke without fire, of course (I'll discuss the "fire" later in this chapter), but there's another way to look at technical writing. For all the complaints about technical writing, most technical writers stick around. I don't know why the jokes circulate—maybe it's simply black humor common to all professions, or maybe it's just the people I hang around with—but if the Society for Technical Communication (STC) is right, most writers are quite happy. In a 1988 survey the society found that 87 percent of its members were "very satisfied" or "satisfied" with the profession, while only 10 percent were "dissatisfied." And the money can be good, too. But before we get to that, let's talk about what technical writers actually do.

Technical writers *explain things*. What sort of things? They explain how to use VCRs, how to use computer programs, how to install telecommunications switches, how to operate blood-testing instruments. Sometimes the technology they are explaining is very complex; it may take a team of sixty writers several years to document a telecommunications switch, for example. Other times what is being explained is fairly simple: how to operate an iron, for instance, or a telephone. Technical writers produce instruction books, user guides, reference manuals, instruction sheets, and so on. But sometimes they "migrate" to slightly different tasks. They might produce public relations brochures, company reports, or even advertisements, or help prepare articles for professional journals. They sometimes end up designing computer program "user interfaces," or even computer-based training courses. Sometimes they write *proposals*—documents used by a company to sell its products. There is an enormous range of documents produced by technical writers. That's a great advantage; it means there are many different ways to enter the profession, and many different routes to take once you're in.

So how much *do* technical writers make? From $10,000 a year to $250,000, sometimes more. That's not a very useful range, so let me elaborate. Let's start by looking at the 1990 Salary Review published by the Society for Technical Communication.

The median salary for a technical writer in the United States, according to this survey, is $35,000. (Canadian writers generally make a little bit less than their colleagues in the United States; their median is $34,000, in $US.) Let's look at the median salary related to years of experience.

Less than two years	$25,600
Three to five years	$32,800
Six to ten years	$37,000
Eleven years and more	$42,000

To me those figures sound a little low. In the Dallas area—where I currently work—many writers with six to ten years' experience are making $43,000 to $50,000. Of course the location does have an effect: The median salary ranges from $30,100 in the Montana to Wisconsin region (the 5xxxx zip code area), to $36,200 in New England (the 0xxxx area).

California/Oregon/Washington is almost as high, at $36,000. Of course, breaking it down by the first number of the zip code doesn't give a perfect picture. Rates in Milwaukee, WI, are probably higher than in Billings, MT, though they are both in the same zip code area. In general, though, rates are higher on the coasts. And even in low-rate areas there are probably pockets of high rates. The 7xxxx zip code area is slightly lower than the national median, at $34,100, but the telecommunication and computer businesses in the Dallas area probably push the rates a little higher. We'll discuss Dallas in more detail in a moment.

This book is about *freelancing*. How much do *freelance* technical writers make? According to the STC survey the average gross income for a consultant/independent contractor is $50,300, and 25 percent are making $58,000 or more. These are the percentages of freelancers in several hourly-rate groups:

Less than $20 per hour	6%	$50 to $60 per hour	7%
$20 to $29 per hour	22%	Over $60 per hour	4%
$30 to $39 per hour	40%	No set fee	3%
$40 to $49 per hour	18%		

So 72 percent of the freelancers surveyed were making over $30 an hour, and 32 percent over $40 an hour. Some of these numbers seem a little *high* for Dallas. "Consultant/

independent contractor" covers a lot of ground: people working through technical service agencies, people working on hourly-rate contracts, genuine consultants, and so on. In Dallas most freelancers I know work through agencies, and most of the experienced ones are in the $27 to $31 per hour range. A few are below—mainly people who are very inexperienced or who don't understand the market. And a few are above; the highest agency rate I have heard of in Dallas is $38 an hour. As for independents, most seem to be in the $38 to $42 per hour range, although those charging by the project—instead of an hourly rate—can often make a lot more ($50 to $75 an hour or more). Still, I've heard rates are much higher in New England and California, so maybe the figures reflect that.

One thing I should note, by the way, is that unlike consultants in many businesses—who spend a lot of time marketing and a few hours working—technical-writing contractors often work at the same contract, full-time, for months or years at a time. So a contractor may be making $40 per hour for 2,000 hours a year, not just 1,000 or 1,500.

A small salary survey published in Consultants' and Contractors' Publications *Job Express* in October 1990 found that technical writers were being sold by the agencies for about $425 a day—$53 an hour—in the New Jersey area. The agencies in turn usually pay their writers $32 to $40 an hour, according to this survey. The $53 figure is important though, because it provides one indication of the market's potential: how much an *independent* writer can charge. (Sure, some clients don't want to pay an individual as much as they pay an agency, but many *will*, if you sell yourself properly.)

According to Bill Oliver of Techwrights, Inc., agencies in New York are paying writers up to $60 per hour (the agencies are making about $65 to $85 per hour), with rates in New Jersey about 15 or 20 percent lower. Writers on long term-contracts in some companies in New Jersey make between $30 and $50 an hour. The higher-paid people are usually Ph.Ds or have a lot of experience.

Now let's take a look at a salary survey published in the May 1991 issue of *Technically Write*, the newsletter published by the Dallas/Fort Worth chapter of the Society for Technical Communication. The survey found that in the Dallas/Fort Worth area of Texas the median annual gross income is $40,000 ($38,000 for employees and $47,000 for contractors; 27 percent of those surveyed were contractors). Here's the breakdown by years of experience (for all writers, both employees and contractors).

0 to 2 years	$23,000	11 to 12 years	$43,000
3 to 4 years	$31,000	13 to 14 years	$48,000
5 to 6 years	$33,000	15 to 16 years	$61,000
7 to 8 years	$36,000	17+ years	$46,000
9 to 10 years	$43,000		

Remember, these are the *median* figures, so half the respondents were above, half below. These figures still look low to me; I've mainly worked in the telecommunications industry for the past few years, though; maybe telecom companies pay more than most companies. These are median incomes, of course, so many writers are making considerably less. Many are making considerably more, too, which is what this book is about: changing the way you sell your skills so you can move up an income group or two.

Incidentally, both the STC and the Dallas chapter report that most of their respondents—and most of their members—are women. Over 57 percent of the Dallas respondents were women, and almost 62 percent of the respondents to the national survey were women. Does this mean most technical writers are women? I don't think so. Most of the writers I know are men, though many of the *younger* writers seem to be women. In fact, the Dallas survey found that the average age of the men was forty-five, while that of the women was thirty-seven. The national survey found a "larger number of women in the lower-paying, entry-level range."

It seems that a high proportion of the newcomers are women, and I have a theory that young women are more likely to join the STC than older men. It may be that while men are more likely to use the "old-boy" network to find work, women entering a new profession seek out professional organizations to help them. Anyway, the fact that there are so many young (women) writers in the STC would tend to skew the median incomes a little, making the numbers look lower than they are in the real world. Do women writers get paid less than the men? Yes and no; in general yes, because they are more likely to be in entry-level positions. But the STC survey found that when salaries are broken down by experience and age, women get paid about the same; in some categories they are paid less, in some they are paid more. For example, women writers with less than two years' experience seem to be paid a little more than the men, as are those with over eleven years' experience, while in between they appear to be paid a little less. But the differences are small, ranging 4 percent above and below the male figures. Technical writing is a profession in which women are doing well, taking many management positions and, it seems to me—a mere male, admittedly—competing on an equal footing with men.

Now, these surveys don't show the highest rates, of course, because they get merged in with all the other numbers. I called the Dallas chapter of the STC to find out what the range was, and was told that there were several "six-figure incomes." At the top was a man in his mid-thirties—with eleven years in technical writing—making $110,000. Next were two more men, each making $100,000, one in his mid-sixties and one in his mid-fifties, both with a lot of experience. All three were independents, of course. (To be fair, I should mention that at the *other* end of the range were two people who said they made only $12,000, and neither was entry-level; I can only assume these people were unable to work—because of sickness or inability to find a job—or didn't need to work.)

But the incomes keep climbing. For example, *The American Almanac of Jobs and Salaries* (1987) talked about a technical writer making $250,000 a year in royalties from books published by a computer-book company named Dilithium, and reported that Mitchell Waite was making $200,000 a year—though Waite was no longer just a writer, as he had started a company that published computer books. And take a look at a SYBEX Computer Books' catalog. As you flick through you will notice little signs saying "75,000 SOLD," "OVER 75,000," "OVER 125,000," and even "OVER 500,000." Now, I don't know what sort of deal these authors had, but I do know that I earned $1.24 a copy from my first SYBEX book. And some of SYBEX's authors have a dozen or more books in the catalog. In one recent catalog Alan Simpson had twenty books listed, with signs indicating that just six of his books had total sales of almost 1.5 million copies. (Simpson has published forty computer books in the last ten years, including multiple revisions of the same book.)

These are not average technical writers, of course; very few reach these heights. In fact, some of the really high incomes are earned by writers who entered the computer book market at just the right time. When they began writing there were few other computer book authors, few computer book publishers, but a rapidly growing market. While some of the early computer book authors became millionaires, such success is far more difficult today. Still, one computer book agent told me recently that his agency had "lots" of writers making $100,000 a year (but lots more writing part-time and making only $20,000).

But great potential still exists; technical writing does not have to be a dead end. I remember one writer telling me he was leaving technical writing to take a sales job. "In sales, your income is unlimited," he told me. Maybe so, but he eventually decided to stay. He discovered that he could make much more as a contract technical writer than in most sales jobs. What he probably doesn't realize is that although there are some sales jobs that pay six-figure incomes, there are also thousands of technical writers around the country making $100,000 or more, and some making over $250,000. And a few—just a few— making seven figures.

Now for the fire behind the smoke that I mentioned earlier. There are many problems in technical writing (though most can be cured by moving to another company—one of the beauties of freelancing). It is true that technical writing can be very boring; *some* technical writing is boring, that is, while some can be very interesting. Some jobs are very dry. One job I did for a telecommunications company consisted of "translating" computer printouts—pages and pages of numbers—into words. It was probably the most boring job I've ever had, almost matched by the time I worked for a Japanese telecommunications company translating "Jinglish" into English. But I've also had some very interesting jobs, like the "windows application builder" I'm documenting right now, a program that uses simple commands to build programs for Microsoft Windows. If you like playing with computers, there's a job waiting for you somewhere in technical writing; you get paid to play with the computer program and then write about it. Yes, some technical writing jobs are boring—especially in telecommunications—just as some jobs in *any* profession are boring. The secret is being able to pick the jobs you want, something you can do after gaining experience and building a reputation.

Another oft-mentioned problem is the lack of respect accorded to technical writers. Again, some companies respect their writers, some don't. There are three reasons for this lack of respect, though. The first is that many companies don't like the idea of paying for documentation. Products are becoming intrinsically more complicated: CD players and VCRs are more difficult to use than gramophones, today's telephones are more complicated than those of just ten years ago, and the rise of the personal computer has created a whole new field of "stuff" to learn. In the past, documentation could be written quickly and easily—and cheaply. Now it takes time and lots of money, and many companies resent this. Some companies, however, have learned that documentation is as much of the product these days as the hardware (or software) itself; after all, if the buyer can't *use* the product, it's no good. In fact the 1988 STC survey I mentioned earlier found that 60 percent of its members believe their employers are placing more importance on technical writing than they had previously. But many companies are still stuck in the old way of thinking and look on documentation as a necessary evil, to be produced as quickly and cheaply as possible.

Second, many managers don't respect their writers because they don't realize how difficult it can be to write a manual; just getting the information out of the developers can be like pulling teeth from a pit bull. Other companies do realize how difficult writing can be. I'm told that American Airlines estimates it takes a whole day to document the information on just one of its software "screens," and that IBM expects its writers to produce half a page a day (though they expect one page a day from a contractor). (If you are working for AA or IBM and your boss expects more, sorry! These numbers come from offices in the Dallas area, and maybe your boss feels different.)

The last reason for lack of respect can be thrown back at the writers themselves. Sad to say, standards in technical writing are very low. I've heard it said that technical writing is one of the few writing jobs in which the ability to write is not really essential—it helps, but it's not absolutely required. That's not just my impression, incidentally, but that of many other writers and managers I have spoken to. After *Software Maintenance News* published an article I had written about the problems with technical writers, I received a call from a computer training company in New York. "We've had so many problems with our writers," the manager said. "I could have written that article myself!" In fact, to save me rewriting, here's that article.

The Trouble with Technical Writers . . .

In 1988 I had the privilege of witnessing a "literary" disaster at a small telecommunications company (okay, so I used *literary* in the broadest possible sense). This company used the services of an international "consulting" firm; the firm provided six technical writers, at $37 per hour per writer—almost $250,000 by the end of the project.

The result? A *few* good books, but not $250,000 worth. In fact over $100,000 worth of writing time was wasted; much of the time didn't result in anything, and the rest resulted in books that the company had to trash.

Unfortunately, such disasters are not uncommon. Perhaps something similar has happened in *your* company, or maybe you've heard of similar cases, situations in which a company spends a lot of money on its writers, only to receive substandard documents—or nothing at all.

What went wrong at this telecommunications company? The first mistake they made was to place too much trust in the technical services agency (or "consulting firm" as they like to be called). For example, one of the writers really wasn't. "His résumé said he was a writer," the documentation manager told me, "but they exaggerated his experience."

Most agencies rewrite résumés to stress items they know the client will be looking for, and sometimes "putting the best face forward" turns into outright misrepresentation. In this case the agency took some minor writing work the employee had done while at college, and blew it up into a full-scale writing job. That "writer" was the first of the group to go.

Agencies are clearly not to blame for all of technical writing's problems, though; most of the problems lie with the writers themselves. In general, technical writers are either not technical or not writers, and often are neither technical nor writers.

Many writers have little technical experience—they may be graduates of journalism or English, for example. This isn't a problem if the person can learn quickly and pick up the technical knowledge necessary to write about a product, but in most cases such writers write beautiful prose with little substance. It sounds good, but it doesn't contain the needed information.

Three of the writers at the telecom company fitted this mold. One, for example, was a graduate of a technical writing course, but had no real-world technical experience. All three had difficulties understanding the technology with which they were working.

But all the technical knowledge in the world doesn't do any good if the writer is unable to get the information onto paper in a form that is understandable. Many writers have strong technical backgrounds—especially the ex-military writers who are so common in some industries—but many of these people write poorly, and are often used to writing in the stodgy *milspecs* (military specifications) style. (You've seen this type of writing; "see paragraph 2.3.4.1.5-A, Utilizing Functionality/Operationality Testing.") They may include all the information, but users won't want to dig through the garbage to find what they need.

Another major problem with writers is one that is probably common in many occupations; laziness. Maybe that's an unkind word. How about "lack of initiative" or "procrastination"? Writers have a tendency to "shoot the breeze" a lot; this was a serious problem with the group at the telecom company. One writer, for example, worked for about three hours, and talked on the phone or with coworkers the rest of the day. And the rest of the group suffered from the same problem to varying degrees.

Now, everyone needs a break now and again, and I won't pretend that I stare at my terminal all day without a rest, but there's a tendency among writers to slip into what the CEO at *another* telecom company called "a country-club atmosphere." (At *that* company, by the way, a group of eleven contract writers produced, at enormous cost, documents that the company's clients refused to accept.)

This laid-back attitude expresses itself in another way: Many writers avoid research and fact checking. They use only the information they have been given (which is often inaccurate or out of date), and rely on the product's developers to "catch mistakes in the reviews." (As you know, few developers have the time to review technical manuals thoroughly.)

The last major weakness suffered by writers is that of poor organizational skills. More than most work, writing demands sensible organization. A book may have all the required information and may be written by a Pulitzer Prize winner, but if it isn't organized well, it's wasted. How often have you looked in a document and been unable to find something *where you know it should be*? How often have you used (or given up using) a manual that forces you to jump from page to page, section to section, or even to another book? Many writers are unable to organize their thoughts, and unable to put themselves in the reader's place and figure out at which point in a procedure the reader will need certain facts.

Having criticized writers (and risked upsetting friends and colleagues!), let me point out what may have become obvious: that technical writing requires such varied skills that it is actually quite difficult. There are few people who are technically proficient, write well, and can organize their work. (Incidentally, the best writers I know are programmers who can write well, but also the worst writers I know are programmers, the ones who write badly.)

It's always easy to criticize someone else—writers complain about programmers and programmers about writers—but it's not so easy to understand the problems that others face. As one writer friend told me, "You can't criticize another writer's work until you've been in the shoes he wore when he was writing." There's a lot to that, because many writing problems begin with the company, not the writers.

There are unrealistic deadlines, for example, and an unwillingness to provide the writers with the information and resources needed to do the job. And writers often suffer from a lack of respect within their companies (okay, it's often justified, but which came first, the chicken or the egg?), being regarded as a necessary evil, on the same level as the product's packaging or shipping label—you've got to have it, but you want to spend as little as possible. These attitudes lead to low morale and a lackadaisical attitude among a company's writers. (I remember the comment of one product manager to a documentation manager: "My 12-year-old daughter could write these books!")

Don't despair, though: It is possible to find good writers, and it is possible to produce good documentation. It just takes a little effort. Just remember this: Pick your writers carefully, and treat them well!

There's a real advantage to low standards, by the way. It's easy to rise to the top. If you are a good writer, present yourself well, and know how to sell yourself (which is what this book is all about), you can do well in technical writing, because you are way ahead of the competition.

2
How to Get Started in Technical Writing

There's a perennial question in the technical writing world: "Is technical writing a profession or a trade?" (Some would answer, "Neither, it's a racket.") Many in the business like to think of it as a profession, but in many ways it really *isn't*—which is lucky for anyone trying to become a technical writer. Okay, I use the term *profession* elsewhere in this book, but I've used it in a general way. The real point of the question is, How easy is it to get into technical writing?

What is a profession? (This is really not a question of semantics, by the way; bear with me and you will find out why.) My Oxford English Dictionary calls it a "vocation or calling, especially one that involves some branch of advanced learning or science." I suppose technical writing could be called a vocation, but what about the learning? No one really knows what it takes to be a technical writer, what advanced learning or science is required. Yes, there are colleges that teach technical writing, and one day you may need a technical writing degree to enter the profession, but that day is a long way off. I know technical writers who have never even been to college; a 1988 Society for Technical Communication survey found that 12 percent of its members had no degree, and the recent survey by the Lone Star chapter of the STC showed that 18 percent of *its* members had no degree.

A friend suggests that technical writing is in the position that computer programming used to be. A few years ago almost anyone could become a programmer without a programming degree—there weren't any degree courses. People transferred into programming from various professions. But newcomers are now expected to have some kind of degree in computing. Perhaps in ten years new technical writers will need a technical writing degree. Perhaps, but not for a while yet.

Quick, name a profession. Doctors and dentists are professionals, right? Lawyers (jokes aside) are professionals. Psychiatrists and accountants are professionals. All these professions have set standards. You can't decide to be a dentist today and start tomorrow. There are specific courses of learning you must undertake, and tests you must pass. Even real estate salespeople have to pass examinations. But not technical writers. There are no specific requirements to be a technical writer, and if you ask ten technical writers what it takes to be one, you will probably get ten answers (go back and ask again and you'll get ten more).

One reason for this is that the term "technical writing" covers such a lot of ground. Writing a troubleshooting guide for a telephone company's central office switch is very different from writing a user guide for a word processing program, or a maintenance manual for a tape recorder. And there are so many ways to produce technical manuals. Some writers *still* use typewriters, as absurd as that sounds, while others are working on Sun work stations (probably most are working with IBM-compatible PCs). The range of software used to write the books is enormous, and the production methods vary, from photocopying typed pages to professional full-color printing.

If you are a newcomer, the great thing about technical writing is that all this diversity means that there's a good chance someone somewhere will hire you, if you just keep looking. If you have a journalism degree, there's a manager somewhere who used to work on a newspaper and will hire you. If you are fresh out of the military, there's a military writer somewhere who will hire you. If you are an English major or worked as a field technician for a few years, there's someone who will hire you. This doesn't mean just anyone can walk in off the street and get a technical writing job, of course, but it does mean that it is possible for many reasonably educated people to enter the field, if they are prepared to spend a little time gaining some extra skills and then looking for work.

How do most technical writers get into the business? Through the back door—by being in the right place at the right time. Most writers seem to drift into it. A technician is out of work and a friend tells him of a company that needs someone with a technical background to document a new product. A secretary is told to write a user manual because no one else is available. A geologist is temporarily without a project, so his company keeps him busy writing a user guide. These days more writers are actually choosing the career in college, but it's still possible to be in the right place at the right time.

What sort of people are technical writers? I think you can break writers down into groups. Here are the basic groups I've seen:

THE EX-MILITARY

There are many men in technical writing who got there thanks to their military careers. They were writing procedures or teaching while in the military, and often got jobs with military contractors when they left. There are a few *young* men who got into technical writing through the military, but most ex-military writers seem to be middle-aged or older.

THE JOURNALIST

Lots of journalists move over to technical writing. They like the reasonable hours, and figure they've had enough of the poverty that seems to go with most newspaper jobs. The 1988 STC survey showed that 10 percent of its members were ex-journalists.

THE ENGLISH GRADUATE

A lot of people leave college with English degrees and then discover that there are few ways to make a good living with such a degree. Technical writing is an obvious choice. The 1988 STC survey found that 32 percent of its members had English degrees.

THE COLLEGE KID

There's a new breed of technical writer appearing: young people who took technical communication courses at college, as a minor or even a degree course, and make technical writing their first job. There aren't many of them around just yet, but the group is growing, and some companies like to stock up on them. In 1988 the STC found that 10 percent of its members had technical communication degrees, but I don't think that is representative of the profession in general, because I believe younger writers are more likely to be STC members than their older colleagues.

THE SECRETARY

Some secretaries manage to make the leap from typing the boss's letters to writing technical manuals. It's a matter of being at the right place at the right time, and having a little technical ability. It doesn't happen often, but some secretaries have managed to quadruple or quintuple their incomes by becoming technical writers.

THE TECHIE

A lot of people cross over from technical professions to technical writing. There are computer programmers, engineers, geologists (me), chemists, and so on—people who used their technical skills to get work writing.

So how do you get into technical writing? I talked to a few writers and found out how *they* got into the business. This is not just for your amusement, by the way. You will find that in most cases these people don't have anything to show they are technical writers: no degrees, no diplomas, no licenses. No one ever said, "You are now a technical writer." The main thing these people have in common—except for the one "college kid" in the group—is that they simply decided they would become technical writers, in some cases even when they were told they didn't have the skills or education necessary.

Most people get into technical writing by being in the right place at the right time, or making sure they are in the right place at the right time. These examples may give you an idea for how you can use your skills to find a technical writing job, or how you can gain the skills to do so.

PETER, 10 YEARS' TECHNICAL WRITING (THAT'S ME)

I was an oil-field geologist at the wrong time, the early 80s (I was actually a "mud logger," if that term means anything to you). I was promoted to manage my employer's Mexico City office and decided to take a vacation before assuming the position, but when I got back the position had gone! The Mexican government had cancelled our contracts, so I returned to Dallas to the company headquarters, where I was given a make-work position. "Write these manuals for the new computer system we are introducing," they said , "and in six months we'll send you overseas again." I spent the next six *years* writing manuals, training users, and installing and fixing computer equipment. Then I got laid off.

I spent about a year running my own import business, and then working in sales jobs to make ends meet. (Which was great training, by the way. If you want hands-on training in selling your technical writing services, spend a few months selling encyclopedias). At this point, by the way, I didn't realize I was a technical writer. I was an oil-field geologist who had been sidetracked—along with thousands of other people—by the decline in the oil business. I didn't think of myself as a technical writer, that was just one of the things they made me do "until the business got better." But eventually I had to decide what I was going to do with the rest of my life. I sat down and thought for a while and decided that I wanted to be a writer. I had considered a journalism career eight years before, but never quite got round to it, and now, analyzing what I really liked to do, I realized that I should be a writer. But there was a problem. Writing doesn't pay much, I was to be a father in a few months, and I had a mortgage.

How then, to make money writing? I knew journalism didn't pay well, at least in the United States, and didn't feel that I was at a stage in life at which I could start at the bottom of the financial ladder. I took a trip to my local library and started going through the career books. Then I found the *American Almanac of Jobs and Salaries.* Looking up "Writers," I found some interesting stuff. Ghost writers, for example, can make $80,000 for writing one book (if you do Lee Iacocca's biography, that is). That would be a nice direction in which to move, but I didn't think it was practical before the birth of my first baby. Then I noticed the section on technical writers. "Experienced technical writers in the private sector usually earn from $30,000 to $35,000," the book said (this was in 1987). Then they mentioned the authors of best-selling technical books. Daniel McCracken's *Simplified Guide to Fortran Programming* had sold over 300,000 copies, they reported, and Gary Brown's *Simplified Guide to Fortran Programming* had sold about 500,000 copies. And these authors were making $1 to $2 a copy. They went on to discuss technical writers making $250,000 a year, and I realized that I had found a potential profession.

I was lucky, of course, because I *had* done some technical writing. I had produced and maintained a book on a computer system produced by my oil-field employer. It was by no means my full-time job, though—I was also training and installing—and my title was Systems Analyst, not Technical Writer. But I reformatted my résumé to stress the writing and started looking.

I don't want it to sound as if I lied on my résumé, because I don't think I did. I listed all my duties; I didn't try to make it look as if I was a full-time technical writer. I did, however, call myself a Systems Analyst/Technical Writer. Why? Well, I wasn't really a systems analyst. That was a title given to me by a company that didn't know what a systems analyst was, and didn't know what to call me. There's no way I could have found a systems analyst's job with my skills (or lack of them in that case). I had to keep that on my résumé, though, because, well, it *was* my title, wrong or not. But I added Technical Writer because that had been one of my major duties. I had been given a meaningless official job title, and felt I had every right to adjust it to make it more meaningful.

Every résumé stresses *something*. You have to stress the skills you think the prospective employer is looking for; there's nothing wrong with that, it's just common sense. If you are selling a car you don't stress that the steel was mined in Brazil, you concentrate on how the car drives, what it looks like, its economy, and so on. When you are selling yourself you do the same, you tell the employer, "I have the skills you are looking for, and this is why."

I was also lucky in that some of my other experiences would give prospective employers more confidence in my technical abilities, not just my writing. I had trained computer users. I had designed computer-program user interfaces. I had installed and maintained computer equipment. I had used computer equipment in a "real-world" environment.

So, I began looking for technical writing work, mainly in classified ads. Some of those ads were from agencies, some from employers. Luckily the first couple of interviews *didn't* get me work. One was for a medical technology company that would have paid $25,000 a year, another was for a $16 per hour contract. When I discovered the agencies—I had no idea that working through the agencies was so common—I started trying to track down as many as I could. I used the Yellow Pages, the classified ads, and word of mouth (you will learn how to build your own list in a later chapter). The more agencies I spoke to, the more I learned. I discovered, for example, that $16 per hour was way too low.

This is how I found out. After the $16 per hour offer I spoke to another agency. They asked me how much I could work for and I said "around $20 an hour." They said, "Yeah, that sounds okay," so I figured I was asking too little. When I spoke to the next agency I said "in the low $20s," and they said, "That sounds about right." So I knew I was nowhere near the maximum. But I was running out of money fast, and the contracts were taking too long to appear. I finally took a salaried position with an agency for $32,000 a year—though I later discovered they probably would have paid $35,000 if I'd pushed.

So there I was, on a "contract" but paid a salary (you'll learn about that sort of relationship in a later chapter). Making enough to get by, but still receiving calls from the dozens of agencies I had sent my résumé to. It's always easier to negotiate when you don't need a job, and I soon found that contract rates in Dallas were really in the high 20s, sometimes higher. When my first contract ran out seven months later I found another at $27 an hour. And when that one ran out I cut out the middleman and got a direct contract at $38 an hour. So in about a year I had gone from thinking that a salary in the mid-twenties would be okay, to making $38 an hour, and it didn't stop there. A year and a half after that I was making $41 an hour, and then I began a more lucrative way to work—the "per-project" contract. But that's enough about me.

RANDY

"I was in college, studying advertising and marketing. I was almost finished—all I had left was an internship I needed to do, a couple of months in an advertising agency. I left it kinda late. By the time I got around to it, only one place was left, a job with Tandy's advertising department.

"I really didn't fit in. You had to punch a clock, and people who were late more than two or three times in, say, a year, were in real trouble. It was real *bogus*, you know. Of course, I was late five times in the first two weeks. The work was really structured, too. We had a list of words we were supposed to use: We could write "sale," "reduced," and "X percentage off." but not much else. After a couple of weeks I'd had enough.

"I called my professor and told him I was going to quit and get a job; I didn't need the degree that bad, I'd just tell people that I had almost finished my degree. He calmed me down, though, and told me to wait a few days, because he knew someone in Tandy. It turned out he had a friend who ran the Software Technical Writing group. I'd never heard of this—this was in 1982, right at the beginning of personal computing, really. Anyway, she offered me a job, so long as I agreed to leave the marketing crap behind; she didn't want any of that garbage in her manuals.

"I worked there for a couple of years, and then tried to escape. A couple of colleagues and I started a rock band, the Urban Hearts (we later changed the name to Damaged Goods, figuring if we ever got famous our jealous competition would have called us the Urban Farts). I was very much influenced by Neil Young. I kinda liked his bohemian lifestyle. Of course I figured out later that it's a lot easier to be bohemian when you've got as much money as he has. Anyway, after a while I went broke and got a job as a technical writer in Denver, and I've been writing ever since."

GEORGE

"I did a two-year engineering course at Caterpillar Tractor, consisting of both classes and on-the-job training. One of my courses was technical writing; I had classes and worked in the service department for a while. When I finished the course I decided to be a technical writer. In engineering you work on just one small part of a product, but with technical writing you can get involved with the whole thing. In technical writing I could use all my skills."

M.

M. went to a small technical institute in Louisiana, to become a computer technician. When she got out of school she found that the course really hadn't qualified her for anything. She worked in the oil business for a while—on the rigs—and then moved to Dallas, where she worked as a secretary for a temporary agency.

On one assignment she worked for a man who had to produce technical reports now and again. She kept finding and fixing mistakes in the reports, so he finally decided that she could do some of the company's technical writing. After a couple of weeks another technical writer, working on contract, suggested that she work with one of the technical service agencies. She got a technical writing contract, and saw her income jump from $8 or $10 an hour to $20 to $25 an hour. A few years later she was finding contracts paying $40 an hour.

CHUCK

"I was a technical intructor in the Air Force. When I got out I discovered technical writing was one of the best jobs offered to people with my skills, so I took a job with the Martin Company (now Martin Marietta), on the Titan missile project."

STEVE

"I was a history major, but I worked with computers for five years. I joined the Army and worked as a Weatherman for three years. I also had to compile documents and produce various reports. When I got out of the Army I had a job as a night-shift supervisor for computer operators. A friend working for Texas Instruments said that with my technical experience TI would take me as a writer, so I tried and got a job."

CARLA

"My father was a famous engineer, at least in my home town. He told me to get an engineering degree or change my name! So I studied civil engineering at Louisiana Tech. I did really badly in all my classes except for technical writing—in which I got an A. My professor tried to recruit me, and I spoke to a few people who told me that yes, you *can* make a living as a technical writer, so I changed my major.

"I did change my name in the end—I married an engineer. I make more money than my husband does, so I think my father's forgiven me."

LOUISE

"I was married at sixteen, so I didn't go to college. I started going at night, doing a business degree. During the day I worked in my small town's newspaper as the gofer—I made the coffee, did the typing, did some reporting, did the layout. I got tired of that and took a job at Harris, while I was still at school, supervising a group of composition specialists. We did page layout, graphics, and so on. But I had to correct the writers' work! They were technical people, and a lot of these technical guys can't write to save their lives. I knew they were earning more than me, and I knew I could do their job, so six months after starting the job I asked my boss to make me a technical writer. He told me he couldn't because I didn't have a writing background. So I changed my major to journalism—the company paid—and they let me start as an "associate" writer. I also took computer courses at college, because I knew I'd need them, and lots of courses at work. They eventually made me a full writer, and I've worked as a technical writer ever since."

CHUCK

"I was taking calculus in college, and it really sucked, but I did real well in English, so I decided against engineering. After a year in college my dad stopped the money and I dropped out. I joined the Air Force as a mechanic, and when I left I got a job at McDonnell Douglas as an editor and later as a writer. I spent seven years there. The most unusual technical writing job I've ever had was for a company that makes hay bailers. My first day on the job was in a small country town in Utah, examining a potato harvester on an airfield near the mountains."

TOM

"I was into poetry when I was young, and joined a small poetry and art magazine in 1972. I got really turned on by magazine *production*: the layout and pasteup and so on. They made me the poetry editor, and then the magazine's editor. Later, when I went to UT Austin, I began an underground black magazine. Eventually I convinced the *Daily Texan*, the university's paper, to do an insert called "Black Print," and I did a few political editorials for the paper. The *Texan* became the first college paper to use computers for production; I helped in the setup and training, and became fascinated with computers.

"Anyway, when I finally left UT with a degree in physiology I found it difficult to get a job. I took a proofreading position at Taylor Publishing, which I hated. Finally I decided I *had* to get out, and started looking for some other way to make money as a writer. I got a job as a technical editor for a hospital software company in March 1979, and then did some of their writing, too. I've been a technical writer ever since. I would really like to find some way to make a living with *creative* writing. But the *problem* with technical writing is that the money is so good. *Writer's Digest* rates it third, after film and TV script writing, in terms of income. How do you replace that?"

MARY

"I was working at Northern Telecom through a technical services agency, doing mainly secretarial stuff really, but some desktop publishing. I was hoping to do some writing and editing, but they didn't let me do much. Now, I had been in advertising for quite a while in the past, so I had done some writing, and wanted to get into tech writing.

"Well, when they closed down our department, the agency moved me back to their offices to do some newsletters, brochures, and advertising. I kept asking them to place me as a tech writer, but they kept stalling. Then they sent me to Xerox as a junior writer, which was good experience. I also went to Richland College and did some of their tech writers' courses. I spent thousands of dollars on those courses, though looking back it was a good investment.

"After the Xerox contract ended the agency really didn't want to help. They had me working in their offices again, and I kept asking for a raise; they were paying me for desktop publishing, but I was writing for them. All they would offer was 50 cents an hour, so I gave them two weeks' notice. I was taking a chance, because I didn't have any work, but at the end of two weeks I offered to continue working for the agency—at a higher rate, of course—until I found a job. So I got my pay raise; it was just a matter of calling their bluff really. Soon after that I got a direct contract at a local company; it was a short contract, but it was nice and I started to build my reputation. I had a little time off without a contract, though I did some work on a seminar for my son-in-law. I bought a computer and work at home now. Right now I have two contracts I'm working on, at pretty good rates.

"When I'm looking for work I call everyone I know. The STC has found me work a couple of times, I think. But I just call people I know, and I've developed a reputation for doing good work, so I can find work by word of mouth."

So how are you going to find a technical writing job? If I were trying to get into the business today, the first thing I would do is join the Society for Technical Communication (you can find their address in Appendix C). The STC has 15,000 members in 125 chapters, mainly in the United States and Canada. Most of the members are technical writers. The STC can help you in several ways, the most important being the instant network it provides. You will meet poeple who hire technical writers, and people who know who is hiring technical writers and where the jobs are. Many chapters even have a "job bank," a listing of current positions and contracts. The STC also runs seminars and sells books on the subject of technical writing, and publishes a journal and a newsletter. You can also ask the members which local colleges run technical writing courses; an STC member quite likely teaches a course somewhere.

Go to the STC meetings, meet the members, and ask the right questions. How can someone with your entry-level skills find a job? Which companies like to hire journalism graduates (or technicians if you are a technician, or military writers if you are ex-military)? Which companies hire a lot of entry-level people? (Some hire mainly entry-level because they are cheap.) Which companies pay well, which pay poorly? Try to get to know the people who actually hire; even if they don't want to hire you now, if they get to know you and you develop your skills, you will find it easier to get hired later.

(By the way, I'm not on the STC payroll; in fact, at the moment I'm writing this I'm not even a member, though my application is in the mail. I *have* been a member in the past, when an employer paid dues—remember to ask *your* employers if they will pay—but let my membership lapse when I got too busy. But the STC is a great way to get to know your colleagues and find work, especially if you are new to the business or new in town.)

Many people will be able to get hired almost instantly. If you are a journalist and you find a manager who likes journalists, you could get a job without any further education. If you are a technician and you find a manager who hates "those language types" and is only interested in people with solid technical backgrounds, you may be in luck. But you can improve your chances a few ways.

Find a local college that runs a technical writing class, and take it (many cities have at least one technical writing class somewhere). Or find out if your present employer runs technical writing classes (some "high-tech" companies have classes for their programmers and technicians). If you don't have any technical skills, take a few courses in computers, or electronics, or telecommunications. Think about the local market first. Are there a lot of telecommunications companies? Computer companies? You *don't* necessarily have to take a *degree* course in technical writing, by the way. If you already have some writing or technical skills you need to fill the gaps in your knowledge, not start over from the beginning.

Take a few classes in word processing and desktop publishing—a lot of companies think of technical writers as highly paid secretaries, and look mainly for "keyboard" skills. Read a few books on technical writing. (See chapter 3 for information on college courses and books.)

Contact SkuppSearch, Inc. (See Appendix B, p. 144.) SkuppSearch finds permanent positions for technical writers, mainly in the New York metropolitan area, but also places some in other areas of the country (and is planning to open an office in California soon). They work with *all* levels, from entry-level to VPs.

Take a good look at your résumé. Does it stress the skills that would help you in a technical writing position? If you are a secretary, stress the computer manual you wrote, the desktop publishing programs you have used, the technical writing and English grammar courses you took. A lot of companies would never consider you—you are not "technical" enough—but someone somewhere *will*, because they are more interested in page layout than technical information. Buy a good book on writing résumés—it's a skill few people have—and clean the garbage out of your résumé. Remove the six months you were out of work, the month you worked delivering newspapers, your high school accomplishments (unless you are young or they really do relate to the job—maybe you edited the school paper, for example). Place the skills that will help you where the person reading your résumé can find them, and place the jobs that are not relevant out of the way—or off the résumé entirely.

Which skills should you stress? These are any technical skills you have; courses in computing; work in data processing; jobs installing or maintaining hardware or software; word processing and desktop publishing programs you have used (the more the merrier); programming skills; systems analyst jobs; user-interface design; books and magazine articles you have written; writing courses you have taken; teaching or company training courses you have run. What about typing speed? Well, maybe, but low down on the list. Some managers *are* interested in typing speed, but most aren't, and if you stress it too highly it looks as if you are just a secretary (which, of course, you may be).

Most important, keep looking for work. Don't be discouraged. Every manager has a different idea of the perfect technical writer; you just have to find the one who's looking for you. And once you have found that first job you become, like magic, a technical writer! The *next* time you are looking for work you will be taken more seriously. Instead of saying, "I'm a journalist, but I'm trying to find a tech writing job," you can say, "I'm a tech writer." *That's* what defines a technical writer. You don't need to pass exams or be certified, you just need to get your first tech writing job.

3
Teach Yourself Technical Writing

*I never write "metropolis" for seven cents
because I can get the same price for "city."*
—MARK TWAIN

How do you learn the technical writing skills you want to sell? There's no room in this book to teach you those skills, but I will give you an idea of what those skills are and where you can get them.

College Courses

You could, of course, take a degree course in technical writing (or technical communication, or business and technical writing, or whatever they happen to call it). Go to your local library and ask to see the college directories. Not all of them are indexed by subject, but these are:

- ARCO's *The Right College* (Prentice-Hall). This lists twenty-four colleges in sixteen states under Technical and Business Writing.
- *Peterson's Four-Year Colleges*; lists seventy colleges under Technical Writing.
- *Peterson's Two-Year Colleges*; lists three colleges under Technical Writing.
- *The College Blue Book: Degrees Offered by College and Subject*; lists thirteen colleges under Technical Communication and fifteen under Technical Writing.
- You could even pay Peterson's College Quest $35 to produce a list of all the colleges that teach technical writing (and match other criteria you specify). You can find the form in *Peterson's Four-Year Colleges*.
- Check your local community colleges. Lots of them now have technical writing or technical communication courses. For example, Richland College near Dallas has a series of about fourteen technical writing courses, ranging from $90 to $250. They cover subjects such as basic technical writing, programming in BASIC, computer graphics, DOS, desktop publishing, publication principles, computer architecture, data communications, and telecommunications. You might even look into correspondence courses; the University of California Extension has a correspondence course, for example. (See Appendix F, Correspondence Course.)

On-the-Job Training

It's often possible to become a technical writer without any specific training. Perhaps you are a journalist and find a manager who believes journalists make the best technical writers, or maybe your company transferred you into a technical writing position because you were in the right place at the right time. If you *have* become a technical writer without any training, it's a good idea to read a few books on the subject, if only to get an idea of what techniques are used in other companies. You might also consider going to a community college course, or see if your company will send you to a technical writing seminar; some large companies even run their own technical writing courses.

Books

There are many good books on technical writing. *Books in Print* lists about 150 books about the subject, and my local library has about thirty. You might want to begin looking at your library. Whether you go to your library or your bookstore, look in these three book categories: Writing, Business Writing, and Computers (many book stores put the books on software documentation in this section). You should definitely get a copy of John Brogan's *Clear Technical Writing* (for prices and publishers, see Appendix E). Brogan's book is about words and sentence structure, removing redundancies, using the active form of verbs, replacing weak verbs, and so on. It has thousands of examples and exercises. Follow the lessons in this book and your writing will become clearer and easier to understand.

You might also read *Technical Writing: A Reader-Centered Approach* by Paul V. Anderson. This 800-page book is a great introduction, explaining how to write a variety of documents, from résumés to hardware manuals. The book covers sentence structure to page layout, visual aids to library research.

I liked *The Technical Writer's Handbook* by Matt Young. This is a dictionary of terminology and misused phrases and words. You might want to read it from start to finish rather than use it as a reference book, though, to help you clear your writing of clutter. I particularly liked Young's humorous writing style.

There are several good books on writing computer documentation. You might start with *The Complete Guide to Writing Software User Manuals* by Brad M. McGehee. For more detail try *How to Write a Computer Manual* by Jonathan Price. This book began life as Apple Corporation's internal guide to writing manuals. It's an excellent book, covering not only writing computer books but also how to test and revise your work, how to review other people's books, and how to recommend changes to the computer program. It has an excellent glossary and pages of checklists.

Writing Effective Software Documentation, by Patricia Williams and Pamela Benson, provides dozens of examples of page layouts—how to lay out a table of contents, a glossary, reference pages, and so on. You could also take a look at *Technical Writing for Business and Industry*, by the same authors.

There are many excellent books on the subject, but spend a few moments reading the book before you buy. You can find many dry, boring tomes that seem to send the wrong message about technical writing: that technical books should be formal almost to the point of being unreadable.

The Society for Technical Communication (STC)

Many of the STC's local chapters organize monthly speeches and lectures and occasional seminars. The STC is also a good source of books about technical writing, and of course publishes several periodicals. Ask STC members for information about college courses and commercial seminars. (See Appendix C for more information about the STC.)

Finally, here are a few general pointers to help you produce useful, readable documents.

1. Before beginning, think about your readers: Who will read your manuals, and what sort of information are they looking for? Think about how the readers will refer to your book, about the sequence of information they will need. For example, many manuals explain computer program features one by one, in the sequence those features appear in the menus or in alphabetical order. That may be fine for a document intended only for reference, but it is very difficult for a new user to work with. Instead, consider putting the information in the book in the same sequence that the users will need it. For example, a book describing a simple spreadsheet program might begin with installation instructions, move on to a "quick start" section (how to open a spreadsheet and enter numbers and simple formulae), and then explain groups of features in the order of probable use—the simple features first, proceeding through the more complicated or infrequently used features to those that few users ever need.

2. Don't use two words when one will do: Research into reading shows that the more concise your writing—with the minimum of unnecessary words—the easier it is to understand. While it is important to ensure that you include all essential information, it is equally important to make sure that your writing is not cluttered with unnecessary words, words that do nothing to increase the reader's understanding. Look at the following two sentences, for example:

Before reading this chapter, use the installation procedure to put the program on your hard disk.

Before reading this chapter, install the program on your hard disk.

The first sentence is five words and ten syllables longer than the second. Not a lot really, but put five extra words in every paragraph of your manual, and your readers will feel as if they are running in knee-deep water. And the first sentence doesn't tell us anything more than the second.

Here are a few examples of redundancies (on the left) and simpler versions (on the right):

Most of the CODP packs . . .	Most CODP packs . . .
The alarm message indicates the alarm . . .	The message shows the alarm . . .
This chapter provides a description . . .	This chapter describes . . .
This manual is intended to explain . . .	This manual explains . . .
Reference should be made to . . .	Refer to . . .
Table 1 gives a listing of . . .	Table 1 lists . . .
The Print Options commands are to be used . . .	Use the Print Option commands . . .
is capable of . . .	can . . .
An alarm message is generated and printed . . .	An alarm message is printed . . .

3. Don't use two syllables when one will do: Say **press** instead of **depress**, **use** instead of **utilize**, **move** instead of **transport**. If the shorter word doesn't say exactly what you mean, use the longer word. But if it *does* say what you mean, why add extra syllables for the reader to dig through? A syllable here and there doesn't make much difference, you might think, but the more unnecessary syllables, the longer it takes to read, and the less is understood.

4. Use clear language, not jargon or "nonwords": Because jargon and nonwords can often be found in dictionaries, many writers think it is acceptable to use them. (Dictionaries are encyclopedias of use, not of correctness. Many incorrectly used words are found in dictionaries because they are in common use.) But words like **functionality** and **irregardless** slow reading and reduce comprehension, and are unnecessary because alternatives are available. And the unnecessary use of jargon will often confuse your readers, especially those with little experience with a product. Why add jargon to computer documentation that may confuse new users? Why say **visual indicator**, for example, when **lamp** will do just as well, or **the octathorp** instead of **the # key**? (Or **user friendly** instead of **easy to use**?)

5. Use the active voice—speak to the reader personally: In "the old days" technical writers wrote as if they were describing the procedures from afar: "The operator presses the Esc key" instead of "Press the Esc key," "The user should open the left panel" instead of

"Open the left panel." Apart from being unnecessary, the detached form of writing leads to more words cluttering the page. Active writing, on the other hand, is direct and easy to understand.

6. Use an eighth-grade reading level: If you follow these rules, your writing will automatically have a low reading level. Researchers usually advise an eighth-grade reading level for technical documentation, not because few people have a higher reading level, but because the higher the reading level, the longer it takes to get the message across. Probably most of your readers have a much higher reading level, and may enjoy reading Shakespeare, Nietzsche, or the *Washington Post* in their spare time, but when they are trying to understand a new product they want the information quickly. While they may want to be challenged by the new product, they *don't* want the documentation to pose an extra challenge.

Many computer grammar-checkers analyze writing for reading level, but if you don't have such a program, try the following method on a few sentences to check your work.

THE FOG INDEX
(1) Add the number of words in paragraph.
(2) Divide this number by the number of sentences in the paragraph.
(3) Find the number of words in the paragraph that have three syllables or more, and divide this number by the number of sentences.
(4) Add the number from step 3 to the number from step 2.
(5) Divide the total by 0.4.

The final number is the approximate grade level required to read the sentence. If you find your writing is way *over* eighth-grade level, go back and use shorter words, remove unnecessary words, and break the sentences down into shorter ones.

7. Break up the text into lots of bite-sized units: Information is more easily absorbed and understood when it is in small bites. Just as it is easier to digest a pound of steak if it has been cut into pieces, it's easier to understand information if it has been divided into manageable pieces! Technical writing "systems," such as Information Mapping™ or Edmond Weiss's two-page modules, are simply ways of breaking the information down into easy-to-understand blocks.

8. Use lots of headings and subheadings: Punctuating a document with lots of headings serves two purposes: It helps to break the text into those bite-sized blocks I just mentioned, and it allows the document to be used as a reference document. The headers let the reader scan through the book, looking for the required information.

9. Use lots of tables and diagrams: Tables make it easier for the reader to find information when scanning through the document, and diagrams help to make the text's explanations clearer. It's easier to show people something than tell them.

10. Use cross-references: Cross-references are often misused. Many writers think that instead of explaining something fully they can just refer the reader to another area of the document, which is often frustrating for the reader. But there *are* many occasions when an explanation will *touch* on a subject explained elsewhere in the book. If that other subject is not essential to the understanding of the current topic, you don't need to explain it again, but a cross-reference should be used to help the reader find the related subject.

11. Include a complete table of contents: Many publications omit subheadings from the table of contents. This is a mistake, because readers often use the table of contents as a form of index. Readers generally refer to the table of contents *before* flipping to the index, because it allows them to pick an area of interest. The more subheadings included in the table of contents, the closer the reader can get to the required information, without referring to the index (an index is usually so detailed that it can actually mislead a reader).

The next time you are in a bookstore, look at the computer books. Almost all include subheadings in the table of contents, and the better ones even provide page numbers for the subheadings.

12. Include a detailed index: If the table of contents doesn't get the reader where he or she wants to go, the next step is the index. There is nothing more frustrating to the reader than an incomplete index.

13. Consider other types of reference aids: There are other tools you can use to guide your users through your book. I like feature tables. Build a table that lists all the commands in your computer program, for example. You can do this in menu order. The table should include a one-sentence description of each command, and can include a reference to the page or chapter that contains a full explanation of the command. Such tables help readers because they fit in with the way most people learn programs; they start the program and then "investigate," opening menus and trying commands, just to see what will happen. The feature or command table explains each of the program's features in simple terms, giving the reader a quick overview of your program. And the table can be used later to help the user find the area of the book that explains a particular command or group of commands.

You also might consider using tables that list all the keyboard commands, display all the available fonts or special characters, or show all the special symbols used by your program. These can be put on the end-pages or inside covers of your book, or even on separate cards. You can even use a table to allow readers to use a reference book as a learning guide. Though the guide may be structured with the commands in alphabetical order, you could add a table that directs the reader through the book in a tutorial sequence, helping him or her learn the program feature by feature, in a logical order.

14. Proofread the document!: This may seem obvious, but many technical writers omit this step, perhaps because it is the most boring part of the writing process. Nonetheless, *nobody* can produce an error-free document (Ernest Hemingway, for example, couldn't spell). You can make the process easier by using an electronic spell-checker (almost all word-processing programs have them now) and an electronic grammar-checker. Combined, these will catch most problems, but never all of them. So read the document after checking it electronically. (Electronic grammar-checkers can only make suggestions, and tend to find "problems" where none exist—only people who understand grammar can use grammar-checkers effectively!)

15. Use a professional editor: Proofreading your own document is important, but so is having someone else read it, preferably a professional editor. Why? It is nearly impossible to catch all the mistakes in your own work. For some reason, writers tend to see what they *thought* they wrote, rather than what they *actually* wrote.

A word of warning about these recommendations: You will find technical writers who disagree strongly with some of them. One company, for example, was quite upset when I replaced the word *depress* with *press* in a book I was editing, and many managers still prefer to use the "formal" or detached mode of writing in their books: "The operator depresses the key," instead of "Press the key." Nonetheless, I believe that most technical writers would agree with these recommendations, in spirit if not in practice.

II.
An Introduction to Freelancing

4
What Is Freelancing?

The word *freelance* comes from the Middle Ages, when it was used to describe a mercenary soldier who would sell himself and his lance to whoever paid the most. Of course its meaning has changed—it no longer refers only to warriors—but the terms of employment remain more or less the same.

The Concise Oxford Dictionary, 7th Ed. says that a *freelance* is a "person working for no fixed employer." My Merriam-Webster dictionary says that a *freelance* or *freelancer* is "one who pursues a profession under no long-term contractual commitments to any one employer." These are good definitions, but you will often hear other terms that tend to complicate the issue a bit.

You will hear the term *contractor*, generally applied to someone who sells his services on short-term contract. You also may hear the term *job-shopper* applied to someone who finds work using the technical service agencies. *Temporary employees* are often freelancers, going from one temporary position to another.

I'm going to use the terms *freelancer* and *contractor* interchangeably in this book to cover all these types of employment, because they all have certain common characteristics. A freelancer moves from one temporary job to another. The advantages are many, perhaps the most important being the ability to make significantly more money. The terms of employment may vary, but the key characteristic is that both the employer and employee intend employment to be of a limited duration. A permanent job comes with an implied promise: "You have a job forever, or until you die, retire, are laid off or fired." The freelance relationship, on the other hand, is recognized to be temporary by both parties: "You've got a job until the work is finished or the contract expires."

You also will hear the term *consultant*, a much misused term defined in one of my dictionaries as "one who gives professional advice or services." Although *consulting* is a form of *freelancing*—and many contractors call themselves consultants—it is important to understand that it is different from mere contracting. A consultant has more responsibility than a contractor; the consultant may have control over the entire project, and may even provide other personnel to do the work. The consultant uses his own methods and techniques, and has control over how and when he does the work. The contractor, however, is often a cog in a machine, doing the work how and when the client says. And an important difference is that contracting jobs are easier to come by than consulting jobs. There are tens of thousands of contract jobs available around the country every day, whereas consulting jobs may not appear until a consultant convinces a client to buy his services. Contractors are the "bread and butter" of the technical services industry, while consultants are the "caviar"; companies know they need bread and butter—so they shop for it every day—but may never buy caviar.

So consultants have to try much harder to find work. They spend much more time looking for work, so although their daily rates are much higher than those charged by contractors, consultants often make no more money. According to a 1988 survey published in Howard Shenson's *Complete Guide to Consulting Success*, the average data processing consultant made $81,102 a year before taxes; but contract programmers can make that or more (one told me he made $125,000 in eleven months), and without the marketing problems that come with consulting. A successful consultant can make a lot of money—most of the technical writers making $100,000 or more are consultants—but there's no guarantee that consulting will automatically increase your income.

Most of this book is about *contracting* rather than consulting. But once you have been contracting for a while you may find that you automatically slip into consulting without any real effort, because the most important method for finding work as a consultant is also the most important method for finding work as a consultant—word of mouth. Most consultants also use other methods to find work, though, such as direct mail, advertising, or writing articles for journals and magazines. I discuss these methods in chapter 22, and tell you how to find more information.

Another term you should understand is *technical service agency*. This is a firm that finds contractors for companies; they charge the client a set amount per hour, pay the contractor an agreed-upon amount, and keep the rest. For example, a client may pay $43 per hour for a technical writer. The agency pays the writer $28 (or $25, or $13, whatever they agreed on), withholds taxes, pays any benefits they promised the contractor, and keeps the rest. The agency may keep anywhere from $5 to $25 of the $43 paid by the client. You also may hear the technical service agencies called *job shops*, *shops*, or just *agencies*.

The term *agency* offends many agencies, incidentally. They like to think of themselves as "consulting companies." I was speaking to a recruiter at a job fair once, when I suddenly recognized her company's name. "Oh, Acme Inc.," I said, "you're an agency, right?" She looked very offended and said, "Certainly not, we are a consulting firm."

"An agency brings parties together for a potential transaction exclusive of the agency itself," wrote the vice president of one agency in reply to an article I wrote in *PD News*. He continued: "A technical contract firm provides services to the marketplace by assigning its own employees to client projects. As long as the contract employee remains on assignment, the technical contract firm is a vital component of the relationship." This VP evidently felt that his company was a "technical contract firm," a subtle distinction that most contractors ignore; as far as I'm concerned, if a company finds a contractor for a client and takes no real part in the project's planning or execution, the company is an agency, not a technical contract firm. (I know that VP's company, and they often provide the bodies without taking any other role in the client's project.)

Finally, many contractors use the terms *captive* or *slave* to refer to nonbelievers, those people who haven't seen the light and gone freelancing but prefer to remain permanently tied to one company.

The Types of Freelancing

There are several ways to freelance:

1. Independent Contract, Hourly Rate: You have a contract with a client to provide your labor or skills for a specifed number of dollars per hour. The client pays you directly, without withholding taxes. Such freelancers are often known as *independent contractors*.

2. Independent Contract, Fee Basis: You have a contract with a client to complete a specified project, for a set number of dollars. Many people working like this call themselves *consultants* or *independent contractors*. You may agree to write a computer program's manual, a sales brochure, or a reference guide, but you are paid the specified fee regardless of how long the job takes.

3. Temporary Employee: You are an employee of the client, but only temporarily. The relationship is understood to be temporary by both parties, and the client generally pays more than he pays his permanent staff. However, the client pays payroll taxes, and withholds your FICA and federal taxes, just as he does with his permanent employees. IBM has a special status it calls *supplemental* employment; you are paid by the hour, get no benefits, have a limited stay, and cannot return to IBM within a certain period (unless they decide to hire you). (IBM offered me a supplemental position for $13.15 per hour while a friend at the same location was making $38 per hour through an agency! But supplementals often get hired by IBM, so you are supposed to feel that the position is a great opportunity!)

4. Contract with a Technical Service Agency: You are paid an hourly rate by a technical service agency, "on a 1099." That is, the agency does not withhold taxes from you—it pays you the entire earned sum, and reports this sum to the Internal Revenue Service on Form 1099. The agency hires you out to a client company. You may be known as a *contractor* or *job-shopper*.

5. Employee of a Technical Service Agency: You are an employee of a technical service agency. The agency hires you out to a client company and pays you an hourly rate, but withholds taxes and treats you as an employee. The agency may even provide some benefits.

6. Salaried Employee of a Technical Service Agency: This form of relationship should usually be avoided. The agency pays what I call a "pseudo-salary." They tell you your salary, then divide the salary by 2080 to come up with an hourly rate. They calculate your pay by multiplying the number of hours worked by the hourly rate. This relationship is often a way for an agency to make a very high profit from an inexperienced freelancer, because they charge the client a high contract rate, but pay the freelancer a low permanent-employment rate. The relationship is especially unfair if the agency is too small to guarantee the freelancer's job when the present contract ends (although some of the larger agencies can do this), because then the freelancer gets a low income without even the security of continued employment. In fact this type of employment is not really freelance employment, and you should usually steer clear of it.

There are probably other types of freelancing, and many freelancers jump from one type to another, depending on what is available. For example, for legal reasons discussed elsewhere in the book, a company may not want to hire an independent contractor. Instead, they could take the freelancer on as a temporary employee. The freelancer's next job may be through a technical service agency, and the one after that may be as an independent contractor charging a set fee.

As you can see, these terms overlap a lot. There is one important distinction to make though, because it affects how you pay taxes, deduct business expenses, and save for retirement: You are either someone's legal employee, or you are an independent business person, what the IRS calls a *sole proprietor*. Where I have needed to make this distinction I have used the term *independent contractor* or *independent freelancer* to refer to a *sole proprietor*, and *agency employee* to refer to someone legally employed by an agency (if your agency withholds taxes, you are an *agency employee*).

Which of the relationships is the most profitable? The independent contractor relationships are potentially the most profitable, because there is no middleman (no agency), and because you can deduct certain business expenses from your taxes. However, this is not always the case. For instance, you may find a company willing to pay you the same hourly rate if you become a temporary employee—and because they also pay your taxes, you end up saving thousands of dollars in social security taxes. Or you may find an agency with a very rich client willing to pay enough money for both you and the agency to do well. (However, if you are an independent contractor with an employment-income pension plan, you may want to avoid becoming someone else's employee, because you won't be able to put that money into your pension. Pension plans are discussed in chapter 14.)

Why Do Companies Use Contractors?

So why do companies use contractors? I sometimes wonder. An obvious reason is for short-term projects; if the company needs someone for only a few months, they have to hire a contractor. But I've heard of contractors working for the same company for eighteen years. That is unusual, but it is common for contractors to stay with one company for a year or even three. Why would a company do that? Many people claim it's cheaper to hire a contractor because the company doesn't have to pay benefits, but this is usually not true. Hiring a contractor is expensive, so why throw that money away?

A common reason is that "the left hand doesn't know what the right hand is doing." A company may have a policy that limits the number of employees in a particular department, but also requires that department to produce an amount or type of work that forces the department to hire someone; the manager may be able to get around this problem by hiring contractors, effectively bypassing the "no new hires" regulation.

Some companies also guarantee their employees permanent employment; they only lay off personnel as a very last resort, and then pay exceptional redundancy benefits. IBM, for example, has this reputation. Such a company may hire long-term contractors if it is not sure that it will continue to require their services once the project is finished.

Weapons manufacturers often hire large numbers of contractors, and many critics claim it is the inefficiency of U.S. Government procurement procedures that leads to extreme waste, from thousand-dollar hammers to contractors on ten-year contracts. If a company simply bills the government for the cost of a contractor, the length of the contract may not worry them, and it's also easier for them to get rid of contractors if the government suddenly stops funding a project.

Small start-up companies are often good sources of contracts. These companies may not want to commit themselves to long-term employees, and may like the flexibility that hiring contractors gives them—it allows them to hire contractors while developing a product, and then release them when the company begins marketing the product and slows down the development phase.

Companies also may hire contractors if they can't find permanent employees to do the job. "It seems that some professions go through phases," one personnel manager told me. "Sometimes everyone in a particular profession wants to work as a contractor, and you just can't find good permanents." That's a problem experienced by Dallas telecommunication companies trying to hire permanent technical writers; it's hard to find good writers, because so many of them have got hooked on high contract rates.

How Much Experience Do You Need?

You may be able to find a contract even if you have little experience; it's not unknown for companies to give contracts to entry-level people. Calling and talking with as many agencies as possible will help you find out. Talk with other contractors as well, but don't let anyone put you off contracting until you've checked every avenue. A *PD News* reader once replied to one of my articles saying that "a new shopper must have a minimum of ten years' experience in his field." He also stated that my articles showed that I was "a successful shopper." Yet I began contracting with only six years' experience, and I know technical writers who began straight out of college.

Of course, it helps to be experienced and skilled—you will find work easier to come by and rates much higher—but if you are a new technical writer you may still be able to find contracts. You can make good money and gain experience at the same time. You certainly don't need to be an expert to work contract. Howard Shenson, talking about consultants, states that "very few are the world's leading authorities in their fields. Instead, they are active, practical, energetic people who put the theory to work and make it pay. You have no reason to feel unqualified just because you do not rank as number one in your field." A friend of mine once told me that he didn't want to become a contractor until he was the best he could be in his field; he felt he had to "grow" more in his profession. Six months later he was contracting at a very good rate, and his client was so happy with his performance that he tried to hire him full-time. So don't feel you are not good enough to be a contractor. As one department manager told me, "Contractors are no different from their permanent-employee peers—they just want to make more money."

The Three-Step Method of Freelancing

Here is a Three-Step Method for becoming a freelancer. In Step One you will contact the technical service agencies. If there is a demand for your skills you can find a contract through an agency, at a higher rate than you now make. You will start to learn a bit about the freelancing market: the kind of rates available, how much work there is in your area, and who is employing freelancers. This step allows you to start saving money—when I went through Step One I was able to save $1,000 a month, and still support a wife and baby. You will begin to strengthen your résumé, working on different projects and products, with different tools and techniques. And you will begin to build a network of contacts; other freelancers, employers, agencies, and colleagues in different disciplines who work on projects that need your skills.

Step One may last a few months, or several years (it took me thirteen months). Or you may never leave Step One—but that's okay too. Many freelancers spend the rest of their careers working through agencies. They find they can remain employed, earn good money, and leave all the sales and marketing to the agencies.

In Step Two, if you decide to continue, you stop using the agencies to find work. You have built a network strong enough to track down the work, and a reputation good enough to get the contracts. You have also saved enough money to survive while you wait for your client to pay your first invoice. You also may need money for business stationery, mail, professional association fees, computer equipment, and so on, depending on how you intend to sell your services. Often the only office equipment you require, though, is a telephone.

In Step Three you will begin selling your services as a consultant, charging by the project rather than by the hour. If you are disciplined and work more quickly than most writers, you can boost your income dramatically, because you base your fees on how much the competition charges; and because your competition is slower than you are, you can make more money for every hour you work than if you bill by the hour. And if you can build a strong reputation for doing good work, your fees can go even higher.

"Money Isn't Everything—What About Job Satisfaction?"

You are going to hear this a lot, mainly from permanent employees trying to rationalize their continued employment: "Sure," they say, "you make good money, but money isn't everything, you know." Of course it's hard to argue with this statement. Job satisfaction is very important. But the statement is based on a couple of premises that may not be true: that freelancers don't have job satisfaction and that permanent employees do.

I worked with a lady who used to tell me job satisfaction was more important than money—when she wasn't complaining about her job. Do most captives have job satisfaction? Probably not. Many are just working till five o'clock; so, if you are just putting in nine-to-five as a way to make money, why not make more money? Okay, some do have job satisfaction, but so do many contractors. I believe the contractors who don't have job satisfaction are often working for companies whose captives don't have job satisfaction either, and for the same reasons—boring projects, patronizing or arrogant management, no respect or evidence of gratitude for a job well done, and so on. But the contractor knows he'll be out of there in a few months, while the captives remain stuck in a rut, until they quit, the company lets them go, or they retire.

Freelancing can provide great satisfaction, not only from the job itself but because of the significant advantages that freelancing brings—as you will see in the next chapter.

Who Can Work Freelance?

This book was written for technical writers, but hundreds of technical professionals can use the same techniques to find work. Anyone whose skills are marketed by the thousands of technical service agencies throughout the country can follow the procedures described in this book to go freelance.

5
The Advantages of Freelancing

You can love your wife, you can love your kids,
you can love your country and you can even love your dog.
But never love a company.
Because no matter how much or how long you love it,
it'll never love you back.
— A LETTER TO *NEWSWEEK*

Why would anyone want to work freelance? Why give up a secure, permanent position for the uncertainties of freelancing? I hear both sides of the story. One afternoon I was chatting with an employee of a company for which I was contracting. "My wife tried contracting," he told me, "but it really wasn't worth the hassle. By the time we figured it all out she wasn't making any more than she would as a full-timer, so she gave up." That evening I was talking with a programmer who had been contracting for about fifteen years. He traveled around the country, from contract to contract, but now wanted to stay in the Dallas area. "I've made $125,000 this year," he told me (it was late in November). "Figure I'll make $140,000 next year."

To a great extent contracting is what you make it. If you know the ropes it can totally change your life. If you don't, it will just be a temporary phase between permanent jobs. Many people who have tried freelancing have not experienced the advantages because they didn't give the life a chance. If you last as a freelancer, these are the advantages you will enjoy.

Variety Is the Spice of Life

For some people, of course, it is the very *sameness* of permanent employment that makes it so repulsive. I just can't imagine having the same job day in day out, year after year after dreadful year. The very sameness of it frightens me. The same people, the same buildings, the same ideas, the same product. A character in a recent movie told his son not to look forward to growing up. "In ten minutes you will be twenty," he said, "and in another ten you will be thirty. After about half an hour you will be fifty, and before you know it you will be retiring. It all goes so fast." And what better way to speed it up than stay where you are, waiting for time to catch you!

No, I need more variety than one job could give me. Changing jobs every six months, or even every year or two, keeps work interesting. New people, new tasks, and new experiences help you maintain interest in your work. And if one gets a bit boring—you'll be in a new one soon anyway!

But there are more tangible reasons for working freelance. Here are some of them.

Money

Skilled freelancers who know how to sell their services can make a lot more money than their permanent-employee colleagues. You may make 50 or 100 percent more on the first job you take, even allowing for lost benefits. By finding a specialty and building a reputation for yourself you may be able to make several times your last salary. Annual incomes of $80,000 are not unusual, and many freelancers make over $100,000.

Get Paid for the Hours You Work

People earning salaries are often expected to work much more than a forty-hour week. If you calculate your hourly rate, you may be surprised how low it is. Working freelance, you get paid by the hour, even if you charge a fixed fee for the project, because you base that fee on your estimate of the time it will take.

Get Paid Now, Not Later

As a permanent employee many of your benefits are in the form of promises:
> "Stay with us and you'll get a week of vacation this year (if we don't lay you off before the end of the year). Eventually we will give you five weeks' vacation each year (if you manage to last fifteen years). You'll get promotions later (as long as you don't fight with your boss). And we have a wonderful pension plan (if the company stays in business long enough for you to collect)."

Moving to a new job can be depressing, because you know you will only get one week of vacation, two if you are lucky, for the first year or two. You know you have to start again building your benefits; waiting for vacation to grow, for your pension plan to become vested, to become eligible for a window office.

Working freelance earns you money up front. No implied promises that are broken as often as they are kept. (How many employees do you know who are trying to stay with a company long enough to get a pension—for the third or fourth time? I can think of a few.) You get a check every week or two, and that's it. No I.O.U. "To Be Paid Sometime in the Future." If you decide to leave the job, you are not throwing away all the "pending" payments.

Time Off

Freelancing lets you plan time off, and take as much as you want. If you want to spend a month skiing each year, do it. One friend sells his services as a technical illustrator for most of the year, and then takes off during the summer so he can play baseball with his sons. Unlike salaried employees, who are limited to one or two weeks' vacation in the first year of a job, you can take as many as you want. And as a freelancer you can not only have more time off, but usually say when you want to take it. Sure, you need to be aware of your client's deadlines, but if you are on a six-month contract, it doesn't usually matter if you take off week 5 or week 15, whereas an employee often has more restrictions on him.

Work for a couple of years and take a year or two off if you want. A contract technical writer I met recently did just that. He sailed his boat from Greece to the Caribbean so he could winter in the islands, and then sailed back to England. It is much easier for a freelancer to take a sabbatical than an employee. When you need your boss's permission, it helps to be your own boss!

A More Balanced View of Life

If you've worked with one company for five, ten, or twenty years, you've got a lot invested (all those implied promises the company hasn't yet fulfilled). Problems at work start to take on a new dimension. An argument with your boss is not just irritating, it may cost you thousands, even tens of thousands, of dollars in lost vacation, savings plans, and pension. That is why you hear about people committing suicide after being laid off, or killing the boss after being fired. The job has become so important that a serious problem becomes insurmountable.

I can hear my friends complaining now: "There's nothing wrong with work," they are saying. "Why shouldn't someone love their profession?" I had better make a distinction between a job and a profession. A profession may be very important to you; it defines who you are and what you do. There's nothing wrong with feeling good about being a doctor or a writer or a programmer. The danger comes when your self-esteem depends on a job rather than a profession. A job is simply the use to which you put your profession. Placing too much importance in a job is placing all your eggs in one basket—someone else's basket!

It seems strange that in the United States of America, a country that prides itself on its citizens' individualism and independence, so few people are truly independent—they allow themselves to be directed and controlled by others in exchange for an illusion called "job security."

Easier to Leave a Bad Job

Because you are not waiting for payments you have been promised, leaving a freelance job is a lot easier than leaving a full-time position. But there is another reason it is easy to leave—you have a different mind-set. You expect to leave eventually anyway, so all you are doing is advancing the date a little (but be careful not to do this too often. You don't want to get a reputation for running out on contracts). I have found that permanent employees are much more wary of losing or leaving a job, because of all that they will lose, but also because they don't feel "comfortable" without a permanent job. "I've always had a job," one laid-off employee told me. "I just don't feel safe without one."

Easier to Find New Work

Finding a contract is often easier than finding full-time work. While it may take several months to find full-time work, it may only take a couple of weeks to find contract work. Why? Probably because companies are more careful about hiring a full-timer—someone whom they want to stay years or even decades—than they are about hiring a contractor who may stay only a few months.

Get a Wide Range of Experience

Working a variety of contracts over a few years can help widen your experience and make it easier to find a job. You will work on a range of products, with different tools and technologies. You will learn more about your profession than you could by staying with the same company for years. Freelancing helps you stack your résumé with skills, making job- or contract-hunting easier.

Travel

You will see later in this book how freelancing can help you travel. Using some of the periodicals that carry advertisements from technical service agencies, you can find work anywhere in the United States, and possibly even overseas. Many freelancers move from job to job, throughout the country, leading a gypsy-type existence.

Use Freelancing as a Stepping-Stone

The money, free time, and flexibility that freelancing gives you can be used in many ways. You can use it to make your life easier and more comfortable right now, or you can use it to fulfill an ambition you have. A programmer friend of mine works freelance to save the money that will allow him to start his own software business. I am using freelancing to get my writing career going; although most of my books are computer books right now, I'm also working on some projects that I hope will move me into a different genre. The time and money make that much easier to do.

You might have found your life sidetracked. After fifteen years as a technical writer, you suddenly discover you would rather be an archaeologist. What do you do? You've got a spouse and three kids, so you can't just drop out and go back to college. But a high hourly rate could help you save money to take the time to study, and to make those long trips through Central America or North Africa.

Make Extra Money on Referrals

Some contractors make extra money by establishing a relationship with a technical service agency and feeding the agency "leads." For example, if the agency needs someone to fill a position, and if you can find someone to take and keep the position for at least a month or two, you can earn a referral fee. Or if you know of a company that needs a contract employee, you tell the agency, and the agency fills the position; then you could get a fee. Not all agencies will do this, but the ones that will typically pay from $500 to $1,500 for each contractor hired or position filled. One technical writer I know made $5,000 in one year from fees.

More Job Security

Contractors have more job security than permanent employees.

Does that statement confuse you? It is my opinion, and that of many other contractors I know, but most permanent employees and perhaps even most contractors think it is nonsense. Contractors, after all, are out of work every year or so, or even every few months. How can you call that job security?

So what is job security? A permanent job doesn't provide it, because you can lose the job—as people do every day. Perhaps Europeans and Australians have job security, because if they lose their jobs they get large redundancy payments, but in the United States such payments are rare; you might get a few weeks' at most.

To me job security means the security of knowing I can find work when I need it, with minimal time out of work. Let me give you an example. I joined a telecommunications company, on a contract with a technical services agency, at the same time another writer joined the company as a permanent employee. He needed to build a pension, he told me, and wanted more job security than contracting could provide. Twenty-one months later he was laid off, spent a few weeks out of work, and had to take a bad job. I, on the other hand, worked contract for two other clients and lost not one day of work.

A freelancer who knows the contract market and knows how to sell his services has more job security than a captive who finds himself thrust into the real world totally unprepared. You may be employed today, but if you are laid off tomorrow what are you going to do? Contractors know the answer to that question; most captives don't.

Forget the Office Politics

Don't office politics make you sick? A real advantage to freelancing is that you don't have to worry about them. So long as the client is paying an acceptable rate, I don't care too much about what is going on around me: who is likely to get the next promotion; who is being fired next; whether the new boss is an s.o.b. or not. I don't even care too much about talk of layoffs and bankruptcy, because I know I'll be leaving soon anyway.

I don't have to play stupid office games to get ahead (compliment the right people, wear the right clothes, say the right things). Companies usually judge freelancers on their work; because they are not seen as company people it doesn't matter if freelancers don't "fit in." And if it all gets too bad, you can always get another contract.

I know that office politics damaged my earlier career. While working in the oil business in Mexico I advanced rapidly, and soon found myself in a management position—just as the oil business started to fall apart. The company transferred me back to Dallas, out of an environment in which performance counted and into one in which image was more important. I had an instant black mark against me—when I joined the company a few years earlier, I had worn an earring.

Earrings on men, as you might imagine, are not common in the oil field, certainly not as common as in the English university from which I'd come. But working on the oil rigs I experienced nothing more than rough, but generally good-humored, mocking. In the world of office politics things were worse. On the rigs people tell you things to your face; in the office they smile while they stab you in the back. It was only through the office grapevine that I heard what managers had said, and discovered why I had lost an important promotion.

No, give me contracting. I don't like playing these silly office games. I don't like having to mold my personality into an image approved by some old man in a little grey suit. I know I'm not alone in this. Most people don't enjoy the silly games. They don't *have* to get involved, if they contract.

Freelancers Like Being Freelance!

Perhaps the best advertisement for freelancing are freelancers who are happy with their careers. The Society for Technical Communication survey we looked at earlier in this book asked freelancers about their work. When asked about their expectations, 55 percent of the contractors and consultants said they felt freelancing was *better* than they had expected—only 7 percent said it was worse. Ninety percent of those surveyed said they prefer freelancing to salaried work, and only 8 percent were actively looking for a salaried position. A 90 percent approval rating is a pretty good endorsement for the freelancing lifestyle.

6
The Disadvantages of Freelancing

There are several disadvantages to freelancing, so you must decide if they outweigh the advantages. I can imagine some situations in which you would be better off not freelancing—if you have serious medical problems, for example.

You Don't Get Any Benefits

Perhaps the most significant problem is that you may not receive the sort of benefits you would normally get from a permanent job. You may have to buy your own medical insurance, for example. (I do not regard vacation as a benefit, by the way—companies do not pay their employees to take vacations, they simply withhold some of the employee's pay until he takes a vacation, to give the employee a constant income throughout the year.)

However, many agencies do have benefits, often very good packages that include medical, dental, long-term disability, and even vision insurance, at very reasonable prices. If you go totally independent, working without the assistance of an agency, you will have to find your own benefits, but as you will see in chapter 14, you can usually replace these benefits by buying your own policies and setting up your own pension plans.

Sometimes, though, you may find that the cost of replacing your benefits is so great that you cannot afford to go freelance. If, for example, you have been with a company long enough to be close to becoming "vested" in a pension plan, you should calculate how much you will lose if you leave—it may be worth waiting a while to get the pension money before you go freelance. If you or a dependent have serious health problems, you may want to consider staying with your present company, or at least only working with agencies that have good health insurance. (Remember, however, that most insurance plans have pre-existing condition clauses, so they may not cover medical problems you already have.)

You Must Have More Savings—Business Capital

In some ways a freelancer needs to be more financially responsible. Your savings are your business capital, and a certain portion should be treated as such. However, freelancers usually have more money to play with anyway, so you can be "irresponsible" with a lot of money and still have some left over to save for your business. And when you become established you may find that work is so easy to get that you don't have to worry too much about money.

Whan I started freelancing I didn't dramatically improve my lifestyle. I did spend a bit more, and take longer vacations, but I didn't rush out and buy a hot tub and a Porsche. Instead, I saved most of the extra income, and used that money as business capital.

You Don't Have Long-Term Work Relationships

Some people want the same work, day in, day out. They don't like change, and prefer to know that they will be working with the same people and doing the same thing two years from now. If this describes you, then don't go into freelancing. Your long-term relationships must occur outside of the workplace if you are a freelancer. That's not to say that you won't make lasting friendships with people you meet on a contract, but generally you will simply move on to the next company and the next set of employees with hardly a thought for the last.

You Have No One to Point You in the Right Direction

Some people like to be given a direction to go in, a set of steps to take that will lead them up a company's career ladder. As a freelancer you are on your own. Either you make your own career ladder, or you forget about professional progress and just work for the money. No one is going to act as your mentor or guide.

You Can't Get Involved in Office Politics

One of the advantages to freelancing is that you don't have to bother yourself with office politics. Yet to some people this is a disadvantage. I know people who have done well in their careers thanks to office politics. These are not incompetent people, but they are not always the best people in their companies, either. They play games such as being the first in the office so that the VP sees them when he comes in (but they spend the first hour drinking coffee and listening to the radio); they forge strong friendships with "mentors," superiors who can give them a shove up the ladder; they subtly criticize their peers in front of their superiors.

These games don't work for a freelancer, because the freelancer isn't going to be around long enough for anyone to care, and the company's "rules of the game" don't apply to them anyway.

You Are a Salesperson

This is one of the major disadvantages of freelancing—you have to sell yourself. How often you have to do so will vary, but one way or another you must persuade someone to buy your services. People who don't mind sales work, or even enjoy it, do well as freelancers. If you can handle calling forty strangers a day looking for a contract, and enjoy going on interviews, then the sales aspect of freelancing shouldn't worry you. But if you have an inordinate fear of rejection, or suffer from what salespeople know as "call reluctance," you may have problems freelancing.

No Established Pension Plan

If you work freelance, you have to find your own pension plan. If you work for a company, you probably will have some kind of plan set up for you. However, as we will discuss in chapter 14, the freelancer has several advantages over the captive worker here. For a start, if you are making more money you can afford to save more.

Feelings of Uncertainty and Insecurity

I believe the major factor stopping many people going freelance is a general feeling of insecurity. People spend most of their lives inside social organizations that provide them with guidelines and rules: first the family, then school, college, and a company. These organizations promise certain things. "As long as you behave yourself," the organizations say, "we'll look after you. We will feed you, clothe you, and provide you with social status and a position in the pecking order. You may not like us much," they sometimes say, "but at least you'll know what to expect."

But going freelance is different. You're on your own, with no one to point the way. Sure, you may have an agency telling you they will find work for you, but you can't bank promises. I know perfectly healthy, skilled, experienced people who have unequivocally stated they feel nervous without a job. If that sounds like you, you have two options: Try freelancing anyway (and hope you will get over the nerves), or continue working as a captive.

More Time Spent Job Hunting

Freelancers generally spend more time looking for work. Whereas the average employee may look for work every three or four years, freelancers may look every six months or year. If you don't like the feeling of being out of work, or of being unsure what you will do when your current job comes to an end, then freelancing may not be for you. But if you know how to find work and have the confidence that comes with knowing that your skills are in demand, then you have no reason to be nervous about freelancing.

You won't necessarily spend more time out of work, though. You will start job-hunting before your contract ends, so you may go straight from one contract to another. One contractor in Dallas told me he has been freelancing for seven years and has had only one week out of work. And I have lost only one week in the past three and a half years.

You Don't Get Vacations

This isn't a disadvantage, but I wanted to discuss it because so many people think it is. "If you work on contract you don't get any vacation," I often hear. What do people think, that I never take time off? "But you don't get paid for your vacation, that's what I mean," people say. Well, I've got a surprise for those people—nor do they.

No company pays people to take vacations—after all, why would they pay you not to work? Companies pay employees to work for them, but they don't pay all the money they owe in each paycheck. They hold some of the money back, and then continue paying the employee while the employee is on vacation.

On the other hand the freelancer receives all his money each paycheck—the client doesn't hold any back for later. I make more money each week than I would if I were a permanent employee, so does it matter if no one is paying me while I'm on vacation? Of course not.

Of course some agencies do give vacation pay, but don't automatically assume that you are onto a "good deal" just because an agency offers you vacation pay—the money still comes out of your hourly rate. I would rather receive the money up front, for a couple of reasons. First, I would like to have the money earning *me* interest, rather than the agency. And second, I would get the money even if the contract ended early—most contracts that pay vacation stipulate that you must work a certain amount of time before you can get the vacation money.

No One Will Train You

Companies usually want to hire people who are already trained. This isn't always true—sometimes they will hire entry-level contractors—but more often than not the company wants someone who can come in and produce as quickly as possible. This may be a disadvantage, especially if you don't have a lot of experience in your line of work. You may decide to remain as a permanent employee for a few years while you learn your trade, and then go freelance. A programmer friend of mine is working for a company that is training him in UNIX. Although he could get more money as a contractor, he has decided not to leave until he is proficient in UNIX.

An alternative is to educate yourself in your free time. My friend could learn UNIX at a local college, for example, although it is difficult to beat the experience that comes from the hands-on use of a technique in the real world.

You Want to Transfer Out of Your Profession

Many people transfer out of the "nitty-gritty" of their professions into management positions. An electrical engineer or programmer, for example, may become a department manager. These avenues for advancement are not normally available to the contractor. A programmer remains a programmer.

You Won't Get Unemployment Pay

If you are out of work, you won't be able to claim unemployment pay. But unemployment generally pays only for up to twenty-six weeks, to a maximum of about $5,000 to $6,000. Successful freelancing will allow you to save enough money to tide you over times of unemployment.

You May Not Be Covered by Workers Compensation

You probably won't be covered by workers compensation insurance, although if you are injured at a client's office you may be covered by their liability insurance. Still, it's important to make sure you have adequate medical and long-term disability insurance. See chapter 14 for information.

You May Not Get Paid

I left this until last because it is not very common. It does happen though, and you should protect yourself as much as you can. If you work for an agency, make sure they pay at least every two weeks—many agencies pay each week. If you work on an independent contract that pays you by the hour, invoice the client at least every two weeks, every week if possible. If it is a small company you should try to arrange favorable payment terms—ten days from the invoice date, for example. You can usually be more confident that large companies will pay, and you may find that most will want to pay within thirty days of the invoice (that's what my contract with one large company said, although they usually paid within seven to ten days).

If you are working for a fixed fee, you should arrange to get some of the fee up front, and the rest at set intervals throughout the project. If the checks are late, find out why immediately, and get the problem sorted out right away; don't let your losses accumulate (if necessary, don't do any more work for the client until the checks arrive, so you don't have too much unpaid time). Incidentally, small agencies are probably more likely not to pay than the larger agencies or client companies.

Using these methods, you can make sure that if a client does default, your loss is limited. Another advantage is that the lower the loss, the easier it is to go to Small Claims Court; you can sue someone in a Texas Small Claims for sums of $2,500 or less. Other states vary, and some have ridiculously low sums. But if a client defaults, check with your local court. Small Claims Courts are cheap and easy to use, tend to favor the "underdog" (individuals up against companies), and may be the only way you will ever see your money.

If you ever need to go to Small Claims Court, get the *Small Claims Court Citizens Legal Manual*. (See Small Claims Court, Appendix F.) This explains exactly how to use this court and what your alternatives are.

7
What Makes a Good Freelancer?

Freelancing is very different from permanent employment, and requires a different temperament. Here are a few characteristics you should have if you want to do well as a contractor.

The Ability to Handle Money

People usually get into freelancing because they can make more money. The extra money allows them to have a higher standard of living—but don't think that if you make an extra $1,000 a month after taxes you can spend an extra $1,000. If you spend everything you earn, and perhaps even more, you might want to avoid freelancing. You are headed for trouble anyway—one day you will find yourself out of a job without any money. Freelancers have to be prepared for more "downtime" than captives (permanent employees). The freelancer needs a certain amount of money in the bank to see him through time between jobs. While a captive may have one job for five years, the freelancer may have ten contracts in that time. If you have skills that are in demand, you may be able to go directly from one contract to another, but it is wise to assume that you will have periods without work, so you need to save money.

Being without money can lead you to take a low-paying contract because that is all that is available at the time. Take the case of a colleague of mine. His first contract increased his income from $13 per hour to $25. The money went to his head. Instead of increasing his spending only slightly, and saving the rest, he started spending much more than he ever had before. In particular, he bought a new car and a living-room suite. He failed to prepare for time out of work.

But everything comes to an end. The department he worked in closed down, and he went from $25 per hour to a $21 per hour contract, all he could find at the time. With money in the bank he would have had time to find a contract that paid at least $28 an hour.

You also need savings for the second phase of your freelancing: for when you get out on your own, independent of the agencies. Although the agencies pay every two weeks, some even every week, independent contractors often wait sixty, eighty, or even ninety days to get paid. You will submit an invoice, which then goes through the company's accounting department, and comes out who knows when. I had to wait about forty days after starting my first independent contract before I received a check.

You must remember that you are in business, and every business requires capital, capital that helps you through the lean times. No business can survive without capital, nor will you if you don't save. I know people working through agencies who would like to work on independent contracts (and earn more money), but are unable to do so because they have no savings. "Six weeks!" said one friend. "I couldn't wait six weeks for money, how would I eat!" This friend is on a good agency rate, high enough to save at least $1,000 a month.

Remember, your savings are an investment in your business. If you can't save, stay out of the business.

The Ability to Handle Uncertainty

I wasn't sure what to call this characteristic, because it is a bit of a paradox. I believe there is more certainty in freelancing than in permanent employment. In freelancing you know that your contract will come to an end, and you prepare for the next contract. Change is not a problem; it is simply one of the conditions of business. All businesses must find a succession of clients, and yours is no different.

In permanent employment, however, you assume your job will go on forever—but it won't. You don't know when it will come to an end, but eventually you will be laid off or fired, become sick of the job and leave, or receive an offer you can't accept ("The company is moving to Gun Barrel, Texas. You've still got a job with us there!").

Sure, I know, some people seem to stay with their company forever, but that is happening less and less these days. High-tech companies, especially, seem to start and crash within a few years, companies often relocate, and people's expectations have become higher—they are more likely to quit to find a better position than they would have a few years ago. ("Depression babies" traditionally hung on to a job as long as they could, but they are retiring now, and their children and grandchildren are more demanding.)

So why do I call this characteristic "The Ability to Handle Uncertainty"? Well, many people *need* a permanent job, even though they are fooling themselves about the permanence. It makes them feel more comfortable, more secure. I've had friends tell me, "I couldn't handle not knowing how long the job will last" as an excuse for not freelancing—and then some of those same friends got laid off from their "permanent" jobs.

Perhaps it is not so much the uncertainty that they object to as the job search—having to search for a client almost continuously. Job searches are a form of sales campaign, and many people do not like what it takes to sell a product, even if the product is their own services. And that leads us into the next characteristic.

The Ability to Sell Yourself

Everyone has to sell themselves at some point. Captives do it when they are looking for a new job, and freelancers do it when they are looking for a new contract. You don't need to be a high-powered salesperson, but it certainly helps.

For example, one technical writer I know moved from Atlanta to Dallas when her husband's company transferred him. Most freelancers would contact agencies in Dallas to find a contract, but she used her sales ability to find an independent contract instead. "I just spoke to everyone I knew in Atlanta, and asked them if they knew anyone in Dallas," she told me. "Eventually I got some names in Dallas and then just started networking." Before she arrived in her new home town she had spoken to four hundred people, and had a contract to go to.

You can avoid most of the selling by working through one or two agencies. But you can be more successful if you sell yourself through many different agencies, or if you sell yourself directly to the client. And if you sell yourself directly you will then get into "cold" telephone calls, sales letters, and even brochures. Successful freelancers are also successful salespeople.

Successful Freelancers Like to Gossip

I could have used a euphemism to describe this characteristic, but "gossip" is more direct and honest. Freelancers who make good money "gossip." Not about people's private lives or scandals (well, not much anyway), but about the business they work in. If they hear about someone getting a new contract, they want to know where and with which company. They want to know how much money people make and how long a contract is likely to last. They ask if the company needs more people, even if they don't need a contract right now.

Eventually you get a feel for the market you are working in. You know where to look for work the next time you need it, and how much each company is likely to pay. You also get to know names of people to call when you need job leads, or the name of the hiring manager in a particular company. Ideally you should even keep a card file, noting the names of other contractors and managers, anyone who may be able to help you the next time you are looking.

Of course you need to be careful when you are talking about how much money people make. Some people may be offended if you ask them this—we are taught that it isn't nice to discuss money, and employers and agencies discourage employees from talking about pay, even threaten them with dismissal. But you need to know how much money people in your line of work earn. You are running a business, remember, and every business has to know how much the competition charges. If you don't know, you may overbid (and lose the contract) or underbid (and lose a lot of money).

I've met people who have been freelancers for years, usually working through agencies, who have no idea how much more money they could ask for. They don't gossip with people much, and so they don't find out the range of rates. For example, in the Dallas area right now technical service agencies normally pay technical writers from about $27 per hour to about $32 per hour. However, some writers are making $15 an hour, and one I know makes $38 an hour. The people making the higher rates are obviously in the minority, and if you don't know what is going on in the technical writing market you might assume that $20 or $22 per hour is a good rate.

So talk to people. Don't make a nuisance of yourself, and don't spend too much of your client's time talking, but do get to know other freelancers, and do get to know the market you are working in.

Be Good at What You Do

If you work through the technical service agencies you don't have to be very good—many agencies fill positions with "warm bodies"—but it helps. As a freelancer you are going to be interviewed more often, have your résumé read more often, and have your references checked more often.

While it may be possible for a mediocre employee to "hide" in a large company (we all know downright incompetent people who somehow manage to hang on to their jobs for years), he will have trouble if he must search for a new contract every six months. If you are good at what you do, though, you can quickly build a reputation for yourself. And a good reputation is worth a lot of money in the freelance business.

Self-Motivation

The freelancer needs to be what my mother used to call a *go-getter*. You cannot just hang around waiting for a contract to arrive—if a contract just "drops in your lap" it probably won't be very good. You have to make things happen, to get up and start looking for work.

Let me give you an example. I have a list of 140 technical service agencies in the Dallas area. I have given this list to several people who either want to work freelance, or already work with technical service agencies. The ideal way to use the list is to contact *all* the agencies at once, as I explain elsewhere in the book. But what do these people do? They call one or two agencies, or mail résumés to ten or twenty, and then accept the first thing that comes along.

If you want to work independent of the agencies, you need to be even more self-motivated. You must be prepared to make twenty to thirty phone calls a day, and keep going till you find something. Most people are not willing to make this sort of effort, so they are better off remaining captive.

The Ability to Handle Change

Freelancers must change gear constantly. One day they are working for a Fortune 100 company, the next they are telemarketing their services, and the next they are working for a small start-up company. To me this is one of the advantages of freelancing, but to many people it is irritating. They prefer to plod along at the same pace, year after year. If that describes you, stay out of freelancing.

The Ability to Learn Quickly

It helps if you can catch on quickly. If you can get rolling in a new project quickly and efficiently you can be more productive, which can only enhance your reputation. A common complaint about freelancers is that they take too long to train—the company has to pay for some totally unproductive time while the freelancer settles in. The shorter this unproductive period, the better you will look.

The Ability to Get On Well with People

As a freelancer your reputation is very important. You go from job to job and work with so many people that you start to get a name for yourself, in a way that doesn't happen to permanent employees. The name you make for yourself can be good or bad, and it comprises two basic components—job skills and human relations skills.

While getting on well with people will not, on its own, get you work, *not* getting on with people can, on its own, lose you work. However well qualified you are, if you have a reputation for being an s.o.b. you will find contract-hunting difficult. On the other hand, if people like you they are willing to forgive a lot. I would rather be known as competent and likable than as a genius but difficult to get on with.

Clearly not all freelancers have all these characteristics. I know some financially irresponsible contractors, some who are not very good at what they do, and others who are no good at sales. Working with the technical service agencies lets you get away with a lot. But the more of these characteristics you have, the more successful you can be, and independent freelancers—those not using the agencies—require just about all of them.

8
How Much Do You Earn?

Before you begin freelancing you must calculate how much your permanent job pays you for each hour you work. That may sound simple; after all, you already know your salary or hourly rate. But your employer pays a lot more than you receive in cash—you have to calculate the value of the benefits you receive. Also, if you receive a salary and do a lot of overtime, your hourly rate may be much less than you imagine.

When I speak of the value of benefits I don't mean how much the benefits cost your employer. I mean how much it will cost you to replace the benefits. The important thing to remember here is that however much a benefit costs your employer, if you don't use it it isn't worth anything. For example, you may have a free membership in a health club, a membership that you have never used. This may cost your boss a couple of hundred dollars a year, and may cost you even more to replace it, but since you don't use it you should not include it in your calculations. We often hear companies and personnel managers explaining how employee benefits add 25 to 30 percent to a salary. Often this is simply not true, but even if your company does spend this amount, so what? When you deduct the cost of the running track you never use and the cafeteria that is too disgusting to go near, you usually find that you can replace the benefits for far less.

You want to know how much it costs to replace your benefits, but the actual value of the benefits—and thus your hourly pay—varies depending on the situation with which you are comparing your permanent employment. For example, it will cost you less to replace your medical insurance if you are working with a technical service agency that has a company-subsidized policy than if you are working with an agency that doesn't. So the amount you earn is a relative number used to compare work situations.

You must calculate how much you get per hour, because that is how you are going to be paid when you are freelancing. If you are working for an agency you will be paid an hourly rate; if you are working as an independent you will either bill for the number of hours worked or charge a set fee that you determine according to an estimate of how long the job will take.

I've seen a few ways to calculate a job's total value. For example, a newspaper in my area advises people to take their base salary, add the value of their benefits, and add a value derived by multiplying the number of days not worked (vacation, sick leave, etc.), times their "daily" salary.

This method is, of course, nonsense. Assigning a value to vacations and adding that value to the salary (which already includes payment for vacations) is ridiculous. After all, if my salary is $30,000 and I take thirty-three days' vacation, how much do I receive? I get $30,000, of course, but my local paper would have me believe that I'm really paid $33,807, because that is the "value" of the time off.

But vacation and sick time have no monetary value. Rather, the more vacation or sick leave you take, the less you work, which means the higher your hourly rate. For example, a job that pays $30,000 a year and provides three weeks' paid time off pays $15.31 per hour. If the same job provided four weeks' paid leave the hourly rate would be $15.63 per hour. (These examples presume a forty-hour work week.) When you assign a value to vacations you are assigning a value to time not worked. So why not assign a value to weekends, or evenings, or lunch hours. . . .

Assigning a value to vacations muddles what should be a simple calculation. The most sensible way to calculate the value of a job (and the *only* way to compare a salary with an hourly rate) is to calculate how much you earn each hour.

When Should You Make These Calculations?

As I said, your hourly income is not an absolute number. Because it includes not only wages or salary but also benefits, and because they vary in value depending on your work situation, your hourly income is in some senses a comparative value.

For example, you earn $15 per hour in wages and benefits excluding medical insurance, and work for 1880 hours a year. What is your total hourly rate? Well, it depends on how much it costs to replace your medical insurance, and that depends on several things. If you are going to add yourself to your spouse's policy for $10 a month, you are earning about $15.06 per hour. If you plan to use an agency policy that will cost you $50 per month, you are paid about $15.32 an hour. But if you are comparing your job with working through an agency that doesn't have a policy, and you estimate that it will cost $500 per month to replace your medical policy, then you are making $18.20 an hour (i.e., you need to make $18.20 an hour to match your present salary and benefits).

So when do you calculate your hourly rate, and what is the point of doing so? I recommend that you first calculate your hourly rate when you decide to work freelance. Some of your figures will have to be estimates. You won't know exactly how much it will cost to replace your medical insurance, but you can guess. Then, when you get your first contract-job offer, calculate again, using the actual figures. And what will the hourly rate tell you? It tells you how much you must earn, in a particular job, to equal your present income.

Let's take an example. An agency has offered you a job. They have medical insurance for $600 per year, but that doesn't include long-term disability, which will cost you $900 per year. You estimate that you can replace other benefits for $1,000 per year. Your present salary is $30,000, and you work 2,000 hours in a year. So your hourly rate at your present job is $16.25.

Salary	$30,000
Insurance	$ 600
Long-Term Disability	$ 900
Other Benefits	$ 1,000
Total	$32,500
Total Hours	2,000
Hourly Rate	$16.25

So what does the figure of $16.25 tell you? It says that working through that agency you must make $16.25 an hour (and be able to work 2,000 hours a year) in order to match your present income. (Remember, however, that many agencies pay time-and-a-half for overtime, which further complicates the issue.)

Now, suppose the agency doesn't have medical insurance, and you estimate it will cost you $300 per month to insure your family:

Salary	$30,000
Insurance	$ 3,600
Long-Term Disability	$ 900
Other Benefits	$ 1,000
Total	$35,500
Total Hours	2,000
Hourly Rate	$17.75

You need to make $17.75 to replace your income.

I have included a work sheet at the end of this chapter. Follow the instructions in the following text to use the work sheet to find out how much you earn.

How Much Are You Paid?

Write down all your income: salary, hourly wages, overtime pay, and bonuses. Include *all* the money, even money paid during vacations and sick leave.

Medical/Dental/Vision Insurance

This one is a bit complicated. If you are married and can get onto the policy provided by your spouse's company, write down that cost. That is probably the cheapest insurance you can get. (If you live with someone in a "common law" relationship you may still be able to get onto their medical insurance, depending on your state's laws.)

Many technical service agencies now have medical insurance (almost all the large ones do). Some charge the full cost to the contractor, but others charge only a portion. One agency I know of charges $50 a month for a company policy, another charges only $10 a month, and yet another charges the full amount ($327 a month). If you have to get your own policy, the cost is going to depend on several things. You can get family coverage for about $300 a month for a small family. However, if you want to include maternity insurance you may need to add $150 a month or more. Also, if you have a history of medical problems you may find insurance very expensive, possibly prohibitively so—you may need to stay in your permanent position just to maintain your insurance.

Another option is to continue your present insurance for eighteen months using a COBRA (Consolidated Omnibus Budget Reconciliation Act) continuance—see chapter 14 for more information. I suggest you examine all your options before filling in this cost. Check a few policies for prices.

If your present employer provides dental and vision insurance and it appears you won't be able to replace it, estimate how much you will lose in a year without it.

Long-Term Disability

You need a long-term disability policy, although many freelancers don't have one. Just imagine the consequences of a sickness that stopped you working for a year or two, or even just a few months. Many agencies include a long-term disability policy with their medical insurance policies, but if they don't, you can replace it (if you are in good health) for around $50 to $100 a month, depending on your age, occupation, and the amount of coverage you want.

Term Life Insurance

Company life insurance rarely has any real value. Although some companies give their employees life insurance with a value of one or two times their annual salary, more commonly companies provide a $10,000 or $20,000 policy. How much is that worth? That depends on your age, sex, and health, but I can increase my own life insurance by $10,000 for about $12 a year (I'm in my early thirties, in good health, and a nonsmoker). Even if your company gives you a large policy, it may be worth only a couple of dollars a week.

For example, if I were a permanent employee I probably would have a salary of $40,000 to $45,000, so my employer might provide a free life insurance policy with a benefit of $80,000; to replace that policy would cost me $95 a year. Also, the tax law changed recently: If your employer provides you with a life insurance policy with a benefit over $50,000, you will now have to pay tax on the cost of the amount over $50,000. For example, if you have a $100,000 policy you will have to pay income tax and FICA on half what your company pays for the policy.

If you don't think your employer's policy is worth replacing (you already have your own or don't need one), don't write down a value.

Tax-Free Savings Plan

Wherever you work, you will have a tax-free savings plan. Even if your company or agency doesn't have one, you can still use an IRA (Independent Retirement Account). However, many agencies also have 401(k) plans; and if you are self-employed you can have an SEP or Keogh plan. These plans let you save even more than an IRA allows. Remember, though, that the value of your present employer's savings plan is limited by how much you manage to save. If your income is too low to allow any savings, the plan has no value to you. If you put less than $2,000 a year into the plan, discount its value, because you can put that amount into an IRA. If you save more, write down the amount that the money saved over $2,000 reduces your tax bill; so if you save $3,000, write down the savings in tax produced by $1,000.

Employer Contributions to Pensions and Tax-Free Savings Plans

It may be difficult to estimate the value of these benefits because they depend on how long you are going to stay with your employer. For many people, especially young people just starting a career, they often have no value because they don't intend to stay with one company long enough to get their hands on the money.

Your employer's contribution to a pension plan doesn't immediately become your property. Typically you must wait over three years to "own" 25 percent, four years to own 50 percent, and five years to own 100 percent. So it's up to you. Decide if you are likely to stay long enough to become "vested"; if so, find out how much your employer contributes (often 3 to 5 percent of base pay for a pension, and a similar sum in matching contributions to a tax-free savings plan). Remember also that if you don't contribute much to your savings plan, your employer isn't going to have much to match.

If you have been with the company a long time these calculations are easier, because you are already "vested," so all your employer's contributions are immediately your property.

Health Club

Does your company provide a membership to a health club? Do you use it? If not, ignore it.

Cafeteria

Does your company provide a cafeteria? Do you use it? How much does it save you each year?

Education

Many companies pay for certain educational courses taken by their employees. If your company is doing this for you, estimate how much you would be spending each year if you had to pay your own way.

Employee Discounts

Does your company produce or sell a product that you use and can buy at a discount cost? If so, do you ever buy any of this product, and would you do so if you didn't have a discount? For example, if IBM lets its employees buy PCs at a 40 percent discount, the cost is still more than the cost of a comparable mail-order clone. Would you buy the product if you didn't get such a discount, or would you be just as happy with the clone? Only include the savings over and above what you would pay if you had no discount. And remember, if you buy one computer every five years, divide the savings by 5.

Day Care

Does your company provide day care while you are at work? If so, how much will it cost you to replace the day care? Don't add that cost to your benefits, though. At the time of this writing you can get a tax credit of between 20 and 30 percent of the first $2,400 you spend on day care for one child or $4,800 for two children. For example, if your adjusted gross income is over $28,001 and you spend more than $2,400 a year for one child's day care, you can deduct $480 off your taxes. So, the value of the benefit is the cost of replacing the benefit, minus the tax credit.

FICA

You don't need to worry about FICA (Social Security tax) if you are going to work for an agency, because usually they will pay FICA for you (and are normally legally obliged to do so). However, if you work as an independent contractor—that is, if you are paid without anyone withholding taxes from you—you must pay FICA.

You will pay FICA on the first $53,400 of your income (1991 figures), and a Medicare tax on the next $71,600. If you are an employee, your company or agency pays 7.65 percent and you pay 7.65 percent of the first $53,400 and 1.45 percent of the next $71,600. But independents pay the full amount, both the "employer's" and "employee's" share, an extra $5,123.30 a year at most. However, that additional sum can then be deducted off your taxes, so you get some of it back! (Aren't taxes wonderful? Who thinks up these systems?) Depending on your tax bracket, deducting $5,123.30 could save you $1,588.22, so the *extra* self-employment tax paid by an independent would only be $3,535.08.

To further complicate things, starting in 1991, self-employed people calculate the income on which FICA must be paid by deducting 7.65 percent from their self-employment income. So if you earn $47,000, you pay self-employment tax on $43,404.50. (Hey, I never said this all made sense.) So an employee earning $47,000 will pay $3,595.50 (7.65 percent of $47,000), while an independent will pay $6,640.89, half of which is deductible. So the independent will probably end up paying about $5,700, around $2,100 more than the employee. ($47,000 minus 7.65 percent equals $43,404.50; multiplied by 15.3 percent equals $6,640.89. Half of $6,640.89 times a tax rate of 28 percent equals $929.73. $6,640.89 minus $929.73 equals $5,711.16. Okay?)

We're going to do some fudging here; we are just trying to come close to an estimate of your pay, so don't worry if this doesn't sound too accurate. If you are an independent and expect your self-employment income minus the 7.65 percent deduction to exceed $53,400, write down $2,940. That amount is approximately how much extra self-employment tax you will pay on $53,400 (the extra 7.65 percent, minus the amount you can then deduct from your income tax—if you understand enough about taxes you can calculate this more accurately yourself, but if not, this should be close).

Then, write down the amount *over* $53,400 that you expect to make, and multiply it by 1 percent. That will show how much extra self-employment tax you will pay on the amount over $53,400 (the extra 1.45 percent, minus the amount you can then deduct from your income tax). If you expect to earn *less* than $53,400, estimate how much you will make, multiply the amount by 5.5 percent (the extra 7.65 percent minus the amount you can deduct from your income tax, about 28 percent), and write down that sum.

You probably don't know right now how much you expect to make, so you might want to do this calculation a few times, for various sums. When you have been contracting a little while you will have more of an idea of how much you can make.

Miscellaneous

There are various benefits that your company may be providing that are very difficult to estimate. For example, you may get funeral leave and moving allowances. If these miscellaneous benefits are likely to be of value to you, add them to the list. However, since you will rarely need these benefits and they are often of very limited value, you probably should just forget them. A company I have done some consulting work for provides some unusual benefits: a free VCR for everyone and a free laser disc player for those who have been employed a few years. Free ice cream and popcorn. Cheap Cokes. The opportunity to smash up a '75 Chevy in the company's parking lot now and again. I have no idea how to calculate the value of such benefits, so if your company provides such things, you are on your own. Remember, though, the key is whether or not you use the benefit. If you don't, it has no value.

Hours Worked per Year

Write down the number of hours you work each year—hours actually worked, not vacation and sick time you were paid for. If you receive a salary, don't just write down 2,080 (52 weeks times 40 hours) unless you actually work 2,080 hours. Subtract the hours you spent out of the office playing golf, on vacation, holidays, sick, taking long lunches, military leave, "personal days," and funeral leave—time you didn't work for which the company paid you. Add all the extra time you may have worked, weekends and evenings.

Work Sheet

HOW MUCH ARE YOU PAID EACH YEAR?

1. Salary/Wages
   _____

2. Overtime Pay
   _____

3. Bonuses
   _____

4. Total Payments—add the values on lines 1, 2, and 3
   _____

BENEFITS—COSTS PER YEAR

5. Medical/Dental/Vision Insurance
   _____

6. Long-Term Disability
   _____

7. Term Life Insurance, Personal
   _____

8. Term Life Insurance, Family
   _____

9. Tax-Free Savings Plan
   _____

10. Employer's Contribution
    _____

11. Company Pension Plan
    _____

12. Company Car
    _____

13. Health Club
    _____

14. Cafeteria
    _____

15. Education
    _____

16. Employee Discounts
    _____

17. Day Care
    _____

18. FICA
    _____

19. Miscellaneous Benefits
    _____

20. Total Benefits—add the values on lines 5 through 19
    _____

21. Total Income (Payments + Benefits)—add lines 4 and 20
    _____

22. Total Number of Hours Worked per Year
    _____

23. Your Hourly Rate—divide line 21 by line 22
    _____

Now you know what you earn. Employees paid hourly are often surprised how high the rate is, and salaried people working long hours are often amazed at how low.

Anyway, now you have a benchmark. You know you must make that much to "break even." You may find you can easily make $10 or $15 an hour above your break-even point, which is $20,000 to $30,000 a year more than you now make. You may be able to increase your income by even more than this. But even if you are not able to make a large increase in your income, I suggest you make sure your first contract is at least several dollars an hour higher than your break-even point. As a freelancer you will be looking for work more often than a permanent employee—you must make more money to see yourself through the times you are selling yourself instead of working.

Many newcomers to freelancing are surprised by hourly rates that sound high. They make $14 an hour in salary, so when an agency offers $21 an hour they jump at it. What they don't realize is that they would receive $18 or $19 including their benefits, so they would increase their incomes only a dollar or two an hour. Make sure you know what you earn and you will know how much a contract must pay to make it worth your while.

III.
The First Step in Freelancing

9
Finding the Technical Service Agencies

Now that you are ready to find contract work, you have to find the agencies. I don't recommend that you work with only one agency. Sure, the first agency you contact may find you something in a few days, but on the other hand they may not, for all their promises. "I tried to work on contract," a captive employee once told me, "but the agency never came through. They kept saying they had something coming up and they would call me, but they never did." He suggested the mistake he had made by using the words "the agency." Had this would-be freelancer contacted a group of agencies, his chance of success would have been far greater.

A few months ago I gave 140 mailing labels to a friend, each one carrying the address of a local technical service agency. My friend, a technical writer, mailed résumés to eighty agencies, and got a poor response. He got about eight responses over a two-week period, eight agencies that had "possible" contracts. The last time I had used a similar list (I mailed résumés to about ninety agencies) I received about twenty responses, and began working on a new contract two weeks from the day I mailed the résumés.

But my friend had such a bad response that I began to feel a little guilty. I had assured him that this was the way to go, that he had to get his name out to as many agencies as possible. He was obviously disappointed. Then I began to consider what would have happened if he hadn't mailed to eighty agencies. What if he had contacted only, say, ten agencies? He may have got only one or two responses.

My friend was looking for work at the end of October, when several local companies had laid off technical writers. The last quarter of the year is supposed to be a bad time to find a contract—companies haven't completed their budgets, people are thinking about Thanksgiving vacations and Christmas shopping and so on. I don't know how much of this is true, but many people who have been in the business long enough to get a feel for the ups and the downs say it is so. You will hear different versions. Some say the troubles begin in November, others say after Thanksgiving, and others even claim that there is a sudden spate of *hiring* right before Thanksgiving followed by a slump. One agent I know claims it is only December that is bad. But most contractors believe the end of the year is bad.

So, here is my friend looking for work in a tough market. All the more reason to contact as many agencies as possible! Yes, his response was low, but he needed as many leads as he could get, and his response would be lower still had he limited his search to a few agencies. Of course he should have mailed to all 140, which would have increased the response. And it didn't cost him much anyway. Sending a two-page résumé and a cover letter to 140 agencies should cost about $60, a minimal expense when compared with the return on your investment.

So let me just run through some of the reasons you should contact all the agencies:

1. MOST AGENCIES DON'T PLACE MANY TECHNICAL WRITERS

Most agencies rarely place technical writers, if ever (though some of the larger ones always seem to have a few writers working). However, that doesn't mean it is difficult for technical writers to find contract work, it just means that they have to work with a lot of agencies. If only one agency in ten has a technical writing position each month, there is no point contacting only five. If your skills are not needed as often as others, you must try harder to find the contracts. Incidentally, there are some agencies that place only, or predominantly, technical writers, though it seems that such agencies don't pay writers very well.

2. SOME AGENCIES HAVE SPECIALTIES

Some agencies have specialties, even if they don't recognize it. Some have intentionally built up business in a particular industry, and others have, just by chance, found most of their clients in one or two types of business. The problem is, though, you can't tell which agency places the most technical writers and which works mainly with airframe engineers. Only if you get your résumé to all the agencies in town are you sure of getting the right ones.

3. IF BUSINESS IS BAD YOU MUST TRY HARDER

If the market is down, you have to find as many leads as possible. You can't do that by calling five agencies and then sitting by the phone.

4. THE MORE OFFERS YOU HAVE, THE MORE CHOICE YOU HAVE

Some people seem to believe that a contract is a contract is a contract, that one is the same as another. Clearly that's not true, but freelancers often act as if the best contract is the first one they are offered. But contracts vary widely. In the past couple of years in Dallas I have met technical writers making $13 an hour and writers making $38 an hour (these are Dallas rates; rates are considerably higher in some areas, much lower in others). One writer I know went from $13 per hour to $25 per hour literally overnight. And contracts don't just vary in monetary rewards either; you may be offered a boring contract, a contract ninety minutes' drive from home, one in a smoke-filled office, or a contract in an industry other than the one in which you want to specialize. When you limit your choices by working with just a few agencies you limit your ability to choose how and where you will work, and for how much.

5. GIVES YOU MORE CONFIDENCE

Imagine you are in urgent need of a job, but you don't feel close to getting a contract. An agency calls and offers you a contract, but at a rate 25 to 50 percent lower than what you believe to be the "going rate." What do you do? Do you say, "No thanks, you're not paying enough," and put down the phone? Do you tell them you can get more money elsewhere and hope they increase the offer? Or do you jump at the chance to start making money again?

Now imagine having a couple of contract offers, plus a few more "in the works." How are you going to feel? A bit more confident, right? You know that if you turn an offer down you've still got other options. The people at the agency will feel your confidence. And it is natural to think that confident people have something to be confident about, that confident people are not worried about being out of work, because they are well qualified and good at what they do. Believe me, you get better offers when you are not in a hurry to find work than when you are desperate to be earning money.

6. MAKE THE AGENCIES COMPETE

Not only will the agencies "feel" your confidence, but you can even tell them that you have other options open. Of course you have to be tactful about this. Don't act arrogantly or imply that they need you more than you need them. Simply mention that you have been discussing contracts with a few other agencies and hope to hear soon about some contracts. In effect you are saying, "If you want me you'd better move, because other people want me, too." Just make sure you don't beat anyone over the head with your ego.

If you are tactful you can play the agencies off against each other. If an agency knows you are close to making a deal with another, the hourly rate they offer you is likely to be higher.

7. CONTINUING JOB OFFERS HELP YOU LEARN THE MARKET

Another great advantage to getting your résumé to as many agencies as possible is that agencies keep résumés on file. I found that if I mailed to ninety agencies I would get one or two calls from agencies each week, for several months afterwards. This helps you plan for the next contract, make useful contacts in the agencies, and get a feel for what is going on in the contract market. When you are working and an agency calls you to see if you need a job, they may try to steal you away from your present contract. When they ask you how much money you are making, exaggerate a dollar or two, and see what sort of counter-offer they make. Listen carefully to gauge their responses. They may say, "Oh, we can beat that" (you are asking too little or just about right); "Well, that's a bit high" (you've got about the right rate and he is just bargaining with you, or you really are asking too much); "That's a pretty good rate" (the same); "That's not bad" (you probably can get a little bit more from them).

Each time you think you have found the top rate, try ratcheting it up a little, a dollar or two; you may find you are a long way below the limit, but no one's going to come out and tell you that. After talking money with several agencies, from a position of strength, you will begin to discover how much money contractors in your profession can make. Naturally you should also be talking to fellow contractors, asking what they make and how much various agencies and companies pay. (But don't be discouraged if a contractor tells you an hourly rate that sounds too low; he may not know the real rates, so just keep asking other people.)

I'm not suggesting that you drop one contract to take another. That is something you should think about very carefully before you do it—you don't want to ruin your reputation. But by talking with the agencies, even when you don't need work, you will learn how much money you can ask for the next time you need a contract.

8. AVOID THE BAD-CONTRACT CYCLE

All these advantages combine to help you avoid a vicious cycle that goes something like this: *i)* you are short of money so you take the first contract that comes along; *ii)* which may not last long and doesn't pay much; *iii)* so the contract comes to an end before you have built up any savings; *iv)* by which time it's Christmas and nobody's hiring; *v)* but you can't afford to wait until January; *vi)* so you take the first contract that comes along; *vii)* which may not last long and doesn't pay much; *viii)* on and on, ad nauseam.

A few contractors will disagree with me on this: Some believe that you should work with only one or two agencies. I received a letter, in response to an article I wrote in *PD News*, from a contractor in Seattle; she told me she had ruined her reputation by working with "several" agencies. Clients got her résumé from so many agencies that they began to be suspicious of her for "not being selective enough" in her job search. I find it hard to imagine how this could happen. While most agencies will check with you before sending your résumé to a client, it's true that some will submit your résumé without talking to you, and occasionally a client will see the résumé from more than one agency. I've had it happen to me, but I don't think it did me any harm, and I doubt that it would happen often enough to ruin someone's reputation. Listen to this, from a 14-year job-shopper, who wrote to me about a *PD News* article: "I had handled four continuous contracts with one excellent technical shop. I became complacent. The local economy changed and the shop moved to another city. I was unemployed for four weeks." He went on to say that he now has a computer database of 264 agencies around the country, and has been out of work for a total of three weeks in the last ten or twelve years.

Finding the Agencies

Your first task is to find all the technical service agencies in your area. I have a list of 140 agencies in the Dallas area, so any large metropolitan area has dozens of agencies. If you have a computer you should make a list of these agencies. Get a database program or a cheap mailing label program (you can buy one for less than ten dollars). If you don't have a computer, just put the addresses into a notebook for now. Leave space to make notes next to each address.

Here is how you find the agencies.

1. ASK COLLEAGUES

Speak to any freelancers you may know. Ask friends and coworkers who work contract or have worked contract, or just know someone who works contract, if they know of any agencies. They may be able to name several, and if you are lucky you may run into someone who has already compiled a list. You also can get good information about the agencies from other freelancers: the agencies that pay well, the ones to avoid, the ones with specialties, and so on.

If a contractor advises you not to work with an agency, try to find out why. He may have a good reason—the agency doesn't pay regularly, for instance, or is so unethical that either you or the client will get burned—but many contractors are too quick to exclude agencies. My list of agencies once fell into the hands of another technical writer, who typed it into her computer and passed it on to other writers. I had no problem with that, but she removed many good agencies from the list.

Many, she said, were personnel agencies (but many personnel agencies also sell contract personnel), and others were disreputable. When I tried to find out what "disreputable" meant, I discovered that contractors had told her they had got a "bad deal" from these agencies, or that they "don't pay very well." That is a strange reason for excluding an agency. How can you possibly know why another contractor got a bad deal? Perhaps the client simply refused to pay any more, forcing the agency to pay a low rate for that job; perhaps the contractor felt he was worth more than the market paid for his services at that time; maybe an aggressive recruiter at the agency managed to convince the contractor to take less than he could have received elsewhere.

One contractor told me that he would "never work for agency X, because of what they had done at client Y." It turned out they had told the client they were paying the contractor $28 per hour, but were only paying $20 per hour. That is unethical, sure, but I had worked for the same agency, at a different client, and with a good hourly rate. Just because they paid this other contractor badly doesn't mean they pay everyone a low rate!

Whenever someone tells me that such-and-such an agency gives bad deals, I usually know of someone else who has worked through the same agency with no complaints. And I often wonder, If the deal was so bad, why did the contractor take it? Because it was better than any other offer he had at the time, of course. You shouldn't be worried about whether the agency gave someone else a bad deal, you are only interested in the deal they will give you. If it's no good, you don't have to take it. Avoid unethical agencies or those with serious financial problems, but don't avoid an agency just because another contractor was not happy with the money they paid him, or you will have no agencies at all.

2. THE YELLOW PAGES

Check the following categories in the Yellow Pages:
- Employment Contractors—Temporary Help
- Computers—Software and Services
- Computers—Systems Designers and Consultants
- Data Systems—Consultants and Designers

These categories include a lot of the agencies you are looking for. Unfortunately they include a lot that are of no use to you. The first category, for example, includes agencies that sell unskilled labor or secretarial help. Some of these "temp" agencies, however, now sell technical services. *Volt*, for example, is a large temporary agency that also places technical professionals, and *Kelly* (of Kelly Girl fame) now has a technical division.

Some of the agency names will tell you that they are no good to you, or that they are probably just what you want. For example, don't bother with *Accountants On Call* or *American Driver Leasing, Inc.*, but add *DocuCorp*. Other agencies have display ads listing the professions they work with. For example, the Dallas Yellow Pages has an ad from an agency called *B & M Associates*; "Experts in contract technical services," the ad says. It lists the types of contractors: "Engineers, Designers, Drafters, Technicians, Computer Science, Programmers, Technical Writers, Artist/Illustrators."

If an agency is clearly of use to you, add it to your list. If it clearly isn't, leave it off. If you are not sure, call and find out. By the way, the Yellow Pages doesn't list zip codes, so you need to get those when you call. If you are trying to build a list of agencies in another town, you can save phone bills by finding zip codes in a zip code directory. You can get a directory from the post office, bookstore, or library.

The Computer categories mentioned include many agencies, but they also include many one-man consulting companies—other freelancers, that is. You will find that out when you call them, although it will be apparent where the business is listed under an individual's name.

3. CONTRACTOR'S PUBLICATIONS

Contact the companies in Appendix A to get sample copies of their publications. They may send a free sample (it will be an old copy, but you can still use it) or they may sell you a copy. Or if you know any contractors, see if they have a copy you could borrow, or check a library. These companies also publish directories, listing 300 to 1,000 agencies; of course these agencies are all over the country, so only a few will be in the city in which you want to work.

Go through these publications looking for agencies in your area. Check all the ads and the lists of agencies, which may be divided into individual states (some agencies may have display ads but may not be listed in the contracts list, so make sure you check all the ads).

4. THE NEWSPAPERS

The job ads in your local papers often include ads from agencies. Check categories such as "Engineering and Technical" and "Data Processing." Some newspapers list jobs by job title rather than the general category, so check a few different, related titles for agencies. For example, if I were checking a California paper I wouldn't just look under "Writer" or "Technical Writer"—I also would look under "Computer Programmer" because I know that agencies using programmers sometimes need writers.

Some companies openly proclaim themselves to be agencies (they usually call themselves "consultants"), but others may have names that imply that they are (*Butler Service*

Group or *Consultants and Designers*, for example, or names with *Associates* at the end). And some agencies can be recognized by the long list of different skills they are searching for. Check the papers every week, because some agencies don't advertise all the time.

Also see Appendix B in this book. This contains a list of national agencies, many of which will have local offices near you. Send your résumé to all of the agencies on this list, or call them and ask for the address of the nearest office.

From these sources you should be able to build a good list. Continue adding to the list whenever you hear of another agency. Once you are working freelance, your contacts should be able to provide you with yet more names.

If you don't live in the area in which you want to work, you can still put together a list. You can use the contractors' magazines, of course, and many libraries stock Yellow Pages from around the country. Libraries also stock newspapers from other cities, but if your local library doesn't, you can write to the newspaper you want (your librarian can help you find the address) and ask about purchasing a copy (many papers have special services for people moving to their cities). If all else fails, you can use a database service (if you have a computer) or take a trip to the other area—after all, if you plan to work there you may as well see what it's like first.

Mailing Labels

Now, what do you do with these addresses? If you have a computer you can use the list to print mailing labels. Most good databases allow you to do this, so if yours doesn't, find one that does (as I said, a mailing label program can cost as little as $10).

If you don't have a computer, buy photocopier labels from a business supplies store. Copier labels are mailing labels on sheets that you can run through a photocopier. You type the address list once, onto blank sheets of paper, and then each time you need to use the list photocopy the addresses onto the copier labels. The labels are self-stick, so all you need to do is lift them off and stick them to the envelopes. You can get boxes of one hundred sheets (33 labels per sheet) for about $14 if you look around. Buy a generic brand; they are much cheaper than the well-known brands, and seem to work just as well.

Preparing the Mailing

A few weeks before you come to the end of a contract or plan to leave your present employer, contact the agencies. Allow three or four weeks, or even do two mailings, one about two months before, and one three weeks before. You need to leave enough time to find the job, but not so much time that you can't take any jobs that come along.

What are you going to mail to the agencies? Not much: a two-page résumé and a one-page cover letter. The cover letter should explain when you are available for work, or how much notice you need to give your present employer, and state that you are looking for contract work (many of the agencies also place full-time personnel, so you want them to know you don't want to work as a "captive"). **Don't** say how much money you want to make, unless you are perfectly sure that the sum is neither too high nor too low. You may want to state the amount later in your freelance career, when you have worked with the agencies for a while, but you probably should avoid it to start with. And I'm not sure there is ever an advantage to stating up front how much you will accept, before you know the details of the job.

I know some people will disagree with me here, but I usually put the résumé and letter on simple white paper, nothing fancy. I don't even personalize the cover letter—I just type

"Dear Sir/Madam" at the top. I suppose it doesn't do any harm to use attractive paper (except for the extra cost, of course), but I'm not so sure it makes any difference to the agencies. After all, when they receive your résumé they are looking for certain professions. If yours is one of the ones they need right now, and it looks as if you have the necessary experience, they will contact you, regardless of the color of your paper. If they are not looking for your profession, they will file your résumé and put you in their database, listed by job title—so again, the color of the paper no longer matters.

What you put on the paper is more important than the paper itself. Limit the résumé to one or two pages (many recruiters say no more than one page). Put your name, address, and phone numbers at the top of the résumé. Robert Marmaduke, with *Apex Technical Services*, suggested in a *PD News* article that you leave three lines between your address and the main text—this allows the job shop to cover your address with their own label when they fax your résumé to a client. Below your address summarize your skills. Remember that you want a recruiter to be able to glance at your résumé and know who you are and what you do. If the recruiter has to dig through extraneous text he might lay your résumé aside before it gets your point across.

Next, list related job experience, with the most recent job first. Don't go into too much detail (you have only a page or two), but include the dates, the company and location, your job title, and a description of what you did. I like to use bold type for the names of the companies I have worked for, which makes the names stand out and catch the eye of the reader. Next, put your formal education, and at the end of the résumé list your special skills. For example, when companies are looking for technical writers they often specify the computer hardware and software with which the user must be familiar. I put a list of hardware followed by a list of software broken down into categories: word processing, desktop publishing, graphics, and so on. There are plenty of books on résumés, so if you are not sure how to put together a good résumé take a trip to your library or bookstore.

Your next step is simply to mail a résumé and letter to every agency on your list. For some reason many freelancers mail the résumés bit by bit—ten today, fifteen on Friday, ten next week. I don't know why they do this, and I can't think of any advantage to doing so. You lose the benefits listed at the beginning of this chapter, because you end up spreading the résumés too thinly—there will be too much time between the contract offers you receive.

Once you've mailed your résumés you can start calling the agencies. This may not be necessary for many people (the agencies will call you), but if times are hard or you don't have a lot of experience, you may need to call and talk to the agencies. This will remind the agencies you are looking and give you an idea of what is going on in the market.

Incidentally, if you don't have a lot of experience you may want to build a relationship with some of the large agencies. Once you have mailed your résumé, call these agencies and talk to the recruiters; try to get an interview and to get to know the people at the agency. You may find that one of these large agencies can help you by including you in a package; clients sometimes hire an agency to provide all the staff for a project, and trust them to find the right people. Some agencies will put inexperienced people in, and may even lie to the client about that person's experience. Not too good for a client, but a good way for you to get the experience! Of course this often backfires—I know of one case where an agency lied and the client found out and fired the contractor. But sometimes it works—the same agency placed the same contractor with another client, where he kept the client perfectly happy.

I'm *not* recommending that you lie, by the way, but as I discuss in chapter 11, many or even most agencies don't check references. That makes it very easy for a contractor to exaggerate his or her qualifications. This often goes wrong for the contractor, because the client discovers the lie or simply finds that the contractor can't do the job, but I'm sure many contractors get away with it.

Mailing Costs

How much will it cost to mail, say, one hundred résumés? Well, the labels cost about 42 cents, assuming you buy a box for $14. The résumés and cover letter should cost about $7.50 (I get my copies made at Bizmart, where it costs 2.5 cents per copy in quantities over one hundred). Generic white envelopes cost as little as 1 cent apiece, so that is $1. The total materials cost is $8.92. Add a little for sales tax, and you can still get the materials for about $10. The stamps cost $29, so it costs about $39 to mail résumés to a hundred agencies. A good contract could earn you that much in the first hour worked.

Life on the Road

Many contractors live a gypsy-like life, travelling from one contract to another. A year in California, six months in Alabama, eighteen months in New England. If you would enjoy working like this (and I must admit, it sounds attractive to me), I strongly recommend that you subscribe to *PD News* or *C.E. Weekly*. These magazines list contracts all over the country, but they also have special services that can be especially useful. Both have résumé mailing services; *PD News* will mail a one-page résumé to all its display advertisers (about forty agencies, although many have several branches) for free, and *C.E. Weekly* will mail a one-page résumé to about a thousand agencies for $240. In both cases you just provide one résumé, and they do all the copying and mailing.

Both publications also have a "hot sheet," a list of contractors seeking work that includes your name, address, phone number, and job description. *PD News* charges $3 to include your listing on a hot sheet that they mail to about 175 agencies. *C.E. Weekly* charges $5 (or nothing if you are out of work) to add your name to a list going to about 1,000 agencies. Or you can buy mailing labels and do your own mailings. (See Appendix A for more information about these services.)

Working with the Agencies

So what happens now that your résumés are in the mail? First, a few of your résumés *won't reach* the agencies. As much as 10 percent, maybe more, will be returned by the Post Office as undeliverable. Companies come and go so quickly that unless you limit yourself solely to companies advertising in a recent newspaper, some of the agencies on your list will be out of business. Even brand-new copies of the Yellow Pages include many out-of-date listings. Just remove these agencies from your list. Don't worry about the wasted postage—that's just part of the cost of doing business.

The next step, we hope, is that you will begin to receive calls from agencies. How many will depend on your experience, the industry you work in, the economy, and the time of year. If you don't get many responses, don't worry—start calling the agencies on your list and talk to them.

Now that you have contacted the agencies, what next? You need to know what to ask them and what to tell them, so turn to the next chapter.

10
Negotiating with the Agencies

Whether an agency contacts you, or you call an agency, there are several questions you should ask before you make any decisions. The agency often won't have the answers the first time you speak to them, but they should be able to answer at some point.

Before I list the questions, let me just explain how the agencies operate. A company's personnel department has one major purpose: to find suitable people to work for the company. The technical service agencies have *two* main jobs: In addition to finding suitable people, they also have to sell their services to client companies. Small agencies usually have the same people doing both things, selling and searching. Large agencies, however, divide these tasks. An agency may have one or two recruiters, searching for potential contractors, and one or two salespeople, looking for potential clients and selling the contractors to them.

When a salesperson discovers a company that needs a contractor, the recruiter has to start looking for candidates. (The salesperson and recruiter may be the same person, remember.) The recruiter will check the résumé file (where your résumé is, waiting for just this moment). He also may run a classified advertisement in the local paper, and call the job banks run by local professional associations. The recruiter then starts calling all the potential candidates, and anyone who may know of a potential candidate. That is when you come in, of course.

When the recruiter calls you, he may tell you that he has a position open. This is close to the truth—he knows a company that has a position open. But it is important to remember that the agency doesn't have a job to offer, however confident the recruiter may sound. It is usually competing against several other agencies, and the client also may be looking for independent contractors. And sometimes the client may not yet have approval to hire anyone. So when you start talking to an agency about a position, don't expect too much, and don't be disappointed if you never hear from some agency again. If the agency doesn't get any further with the position, they probably won't bother to call you back to tell you. (This is very irritating, of course, but it's just the way some agencies do business, and another reason that it is so important to get your résumé out to as many agencies as possible.)

Although you need answers to all the following questions at some point, the first six are the most important—you should ask them immediately. If the answers are not to your liking, there is no point letting the agency submit your résumé.

1. IS THE JOB A PERMANENT POSITION OR CONTRACT?

Some agencies provide both full-time and contract personnel. When you send out your résumé you will be mailing to agencies who work both ways. You also will find, as your network grows, people are hearing about you from other sources. Perhaps a coworker mentioned you to a headhunter, or a headhunter passed on a lead to a friend in another agency. However it happens, you will start receiving calls from people you've never heard of before. Some of those may be employment rather than contract agencies. So if you don't know for sure that the agency you are talking to is a contract agency, ask if the job is contract or permanent.

2. IS THE AGENCY LOOKING FOR CONTRACTORS OR EMPLOYEES?

This is not quite the same as question 1. As I mentioned in chapter 4, some contract agencies only hire people full-time. Even though the job is a contract (i.e., the client is paying the agency an hourly rate to provide people for a limited time), the agency hires people full-time and pays them a salary. Other agencies will employ both full-time people and contractors. Now, it is important to understand that the IRS regards both types of workers as employees.

The agency will withhold taxes from both types of employee, but regards one type (the contractors) as temporary employees and the other type as permanent employees. The permanent employees get the usual employee benefits (paid vacation, medical insurance, and so on) and receive a salary, whereas the contractors are paid by the hour and don't usually get benefits. Of course the line blurs sometimes—some contractors get vacation pay and medical insurance—but the most important distinguishing characteristic is that the permanent employee receives a salary (or a pseudo-salary) while the contractor gets paid by the hour. (See chapter 21 for a discussion of pseudo-salaries.)

For reasons already explained, the best deal is contract, so you don't want to work for a contract agency as a full-time employee. You should bear in mind, though, that even agencies that say they never hire contractors will do so if they really need you. And agencies that encourage you to become an employee will still let you work contract if you insist. (The reason they want you to be an employee is that they make more money, money that could go into *your* bank account!) Anyway, don't let anyone talk you into becoming an employee of their agency, unless it is your only option.

3. WHAT INDUSTRY IS THE JOB IN, AND WHAT TYPE OF WORK?

You may have preferences about the type of work you are looking for, or may have specialized in a certain type of work. For example, there is a lot of work in the telecommunications industry in the Dallas area, so many writers prefer to specialize in that industry. You must decide what you would prefer to do, and what will look good on your résumé the next time you are looking for a contract.

4. FOR HOW LONG IS THE CONTRACT?

The agency may tell you "long term," but who knows what that means? Ask how many months the contract will last. This affects how much your hourly rate should be. If the contract is only a few weeks the rate will usually be—and certainly **should** be—higher than if they expect the job to last eight months. Incidentally, it seems that clients are more likely to extend contracts than to end them early. Sometimes, of course, contracts do end unexpectedly early, but that is just one of the occupational hazards. It is comforting to know, however, that the opposite is usually the case.

5. HOW MUCH IS THE AGENCY PAYING?

Now we get to the nitty-gritty, the most difficult part of the negotiations. When you ask how much the agency is paying, the agent probably will say "Well, how much are you looking for?" Or they may mention a wide range, with the top figure high enough to interest you (later they can say the client wouldn't pay the top rate). They don't want to offer an amount higher than you are willing to work for. I've had agencies tell me $18 an hour, go to $24 an hour when I didn't sound interested, and end up at $40 an hour when I told them I was already on $38 an hour.

You must find out what the current rates are for someone with your experience. You have to do your homework, as explained in chapter 9, and talk to a lot of agencies. Eventually you will get a feel for how much money you can ask for. The first contract you accept may be at a lower rate than you could have got, but if you have done the calculations in chapter 8 you will know how much to ask for to ensure that you work with a smile on your face. And as you talk to other freelancers and contract agencies, you will get an idea of what rate to ask for the next time you need a contract.

Try not to give too much away when you talk rates with an agency. If the rate sounds a lot, don't get excited. As much as it sounds, it may still be lower than you can get. Don't sound as if you need a job soon—the more urgently you need a job, the lower your rate will be. Try to sound confident, in control. Let the agency know you have other "irons in the fire," that they are not the only agency you are talking to. Even if the contract they offer you sounds great, don't let the agency hear the excitement in your voice. Let them think you will carefully consider the contract, not that you'll jump at it.

If you are in a salaried position, don't discuss your present income. If anyone asks, say what I used to say: You are not willing to discuss your present salary, because you know it is less than someone with your skills and experience can earn—which is why you want to work contract. I believe it is unreasonable for agencies to base your rate (or companies to base your salary) on what you **used to** earn. You should be paid the market rate for someone with your skills and experience. If necessary, come up with an excuse not to tell them, or even make up a salary. When I was first looking I told the truth: that my salary was lower than many people with the same experience because for two years my employer had a pay freeze (I worked in the oil business). Therefore, I went on, I didn't want to discuss my salary because I didn't feel it had any bearing on what my hourly rate should be.

Don't be scared to ask for a high rate. Many contractors feel that if they don't take the first position that comes along, it may be a long wait for the next one, so they don't want to jeopardize the contract by asking too much. This may be true for some people (entry-level writers, for example), and some areas of the country—that is something you will have to determine for yourself. Many writers, however, will find mailing résumés to every agency in town will result in dozens of job offers, and don't need to grab the first offer that comes along. Don't be panicked into taking the first job that comes around from fear of losing this one. (Of course, if you are out of work and need a job in a hurry, the picture changes a little. The more you need a job, the less money you are likely to get—so don't let an agency know how badly you need the work!)

Remember, the lower the rate you ask for, the lower the rate you will get.

6. WHAT COMPANY IS THE CONTRACT WITH?

Most agencies won't tell you this until they have an interview set up, but some will, so why not ask anyway? It is important information that has a direct bearing on your decision, so why not find out as early as possible who is hiring?

7. WHERE IS THE CONTRACT? (WHICH CITY OR AREA OF THE CITY?)

Even if the agency won't tell you who the job is with, they should be able to tell you where it is. Remember, a long drive to work each day costs more in gas, wear and tear on your car, and time that could be better spent (working overtime, or playing with your children). For example, you are comparing two potential contracts; one is close to home and pays $25 per hour, the other pays $27 per hour but means an extra hour of travel each day. You would usually pick the lower-paying job. If the first job allows you to work overtime you can work the extra hour and make another $27 a day, as compared with the extra $16 a day you would get from the second job's hourly rate. And you have lower commuting expenses.

8. HOW MUCH OVERTIME IS AVAILABLE? HOW MUCH OVERTIME WILL THE CLIENT EXPECT? IS THE OVERTIME PAID AT THE SAME RATE?

You may not want to work overtime, so you need to know if the client expects you to do so. Some clients may want a job done in a hurry and expect long hours. One client offered me a job that required eighty hours a week. It's a great way to make a quick buck, though it can be depressing after a while.

If you do want to work overtime, you need to know how much they will allow. Many contracts will specify no more than forty hours a week, or may allow overtime only occasionally.

Although most contracts pay the same rate for every hour worked, some contracts may pay time-and-a-half for overtime. This can make overtime very attractive. A high hourly rate, multiplied by 1.5 after the first forty hours, makes it hard to turn down overtime! This is rare—most agencies don't pay overtime rates—but it's worth checking. As you will see in chapter 16, agencies that don't pay may be breaking the Fair Labor Standards Act, but few agencies seem aware of this act.

9. DOES THE AGENCY HAVE A MEDICAL POLICY? HOW MUCH IS IT, AND WHAT DOES IT COVER?

Most agencies can provide a medical insurance policy. If you need insurance you should ask how much it will cost (remember to tell them if you want to include dependents). You will usually pay the full cost of the policy, although some agencies will share the cost with you. For example, I was paying an agency $260 a month for a comprehensive insurance package. It included an excellent medical policy, vision and dental insurance, a small life insurance policy for myself and my wife, and a long-term disability policy. At the same time, however, a friend was paying her agency only $50 a month for a similar policy. However, remember that any benefits the agency pays for come out of what the agency budgets as your cost. While one agency may pay some of the medical policy, another may give a higher hourly rate. In fact, although my friend paid $210 a month less than I did for insurance, she also earned $3 an hour less.

Ask what the insurance covers. There is no need to go into detail until you get close to a contract offer. But eventually you need to know if it covers visits to a doctor, and vision and dental, and find out what the deductible and out-of-pocket expense is. You probably don't need to spend too much time on this the first time the agency calls, but eventually you will need answers. And even if you don't need a medical policy, you might want to ask about their coverage anyway. If the agency doesn't have an insurance policy, or if it has a particularly bad policy, you can use that as a bargaining chip to get a higher hourly rate. (See chapter 14 for more information on medical policies.)

10. DOES THE AGENCY HAVE A LONG-TERM DISABILITY POLICY?

The agency's medical policy often includes a disability policy. You must make sure, though, as it is very important. Many people forget about long-term disability coverage, but being out of work for an extended period can have serious consequences. If the agency doesn't have coverage, you can buy a policy from an insurance agent, but you should use the lack of coverage as a bargaining chip to get a higher hourly rate. (See chapter 14 for more information on long-term disability policies.)

11. DOES THE AGENCY PAY FOR VACATIONS OR SICK LEAVE?

Ask this question and a lot of agencies will sound offended and tell you that this is a contract job, not a full-time job, and that any vacation pay would come out of your hourly rate anyway. They are quite right, but some agencies do pay for these benefits. And yes, it does come out of money that could be on your hourly rate. If I was working for such an agency I would calculate the value of the benefits and ask for the agency to increase my hourly rate correspondingly. "A bird in the hand is worth two in the bush." I would rather have the money now than wait for it, for several reasons. First, I would prefer to have the money earning interest for **me**, rather than the agency. And if for some reason the contract ends early, I would rather walk away having already received my vacation money than have to argue with the agency about whether I'm owed vacation or not. Anyway, it's always good to see an extra dollar or two on your hourly rate—at the very least it means you can tell the next agency that calls that you are on a high hourly rate, and you don't have to lie!

Vacation and sick leave are linked to the number of hours you work, so ask what the ratio is. For example, the agency may pay for forty hours' vacation for every 490 hours you work. In other words, for every 490 hours you work, the agency pays you for 530. Multiply your hourly rate times 530 hours (the time worked plus the vacation time). Now divide the total by 490, the hours worked. For example:

$30 per hour times 530 hours = $15,900
$15,900 divided by 490 hours = $32.45

Then $32.45 is the true hourly rate (presuming you work enough to get the vacation pay). Also, ask when vacation pay becomes "vested." For example, an agency tells you that you must work six months to get six days' vacation pay. Once you have worked your first six months and received the first six days' pay, do you have to work a full six months again to get more vacation pay? If the contract ends five months later, will you get five days' pay or will you get nothing?

In the above example I would ask the agency to pay me $32.45 an hour, instead of giving me vacation pay. However, this negotiation is like buying a car and trading in your old one. Make the best deal you can on your hourly rate **before** you say you want a higher hourly rate instead of vacation time. The increase in your hourly rate should come from the vacation money, not from the dollar or two "negotiating room" they may have left for themselves.

12. DOES THE AGENCY PAY A MILEAGE ALLOWANCE?

Some agencies will pay you driving expenses between your home and the contract. It may not be much per mile, but it all helps. Most contractors and agencies think this mileage allowance is not taxable, but it usually is (see chapter 20 for an explanation of why commuting is not tax deductible.)

13. DOES THE AGENCY HAVE A 401(K) PLAN?

If the agency has a 401(k) pension plan, you will be able to save a larger portion of your income than an IRA allows. However, many 401(k) plans are set up to allow only people who have worked with the company for a year or two to take part.

14. PERSONAL CONSIDERATIONS

You may have other considerations. For example, dress code: Do you want to work at a company that will let you wear jeans? Do you want to work in an office that has banned smoking—or would you prefer to work in a company that still allows smoking? Would you prefer to work in a company with a cafeteria? You should ask these questions later in the process, and you should remember that the more job options you have, the choosier you can be—yet another reason to work with as many agencies as possible.

Road-Shopping: a Few Questions to Ask When Working Out of Town

1. IS THE JOB PAYING A PER DIEM?

Per diem is Latin for "per day." My dictionary says that a per diem is "a payment or allowance for each day." In the contract business, however, per diem means an extra payment for expenses, usually expressed in dollars per week. If the job they are offering is out of town, you should ask if the agency will pay a per diem. If they will, and they tell you how much, make sure you know what they mean. If they say they'll pay a $70 per diem, ask if that is per week or per day (probably per week).

The agency will usually pay the per diem without withholding any tax. Don't think it is not taxable, however. You must keep records of your expenses while you are away from home. If the per diem ends up more than your expenses, Uncle Sam will want a share of the excess. And the per diem may be taxable even if you can prove your expenses. (See chapter 20 for more information.)

2. WHAT ARE THE TAX RATES IN THE AREA YOU ARE GOING TO?

If you are going to work in an area that has a state income tax (most states do), you need to know how much that tax will be. It may be lower than you pay in your home state, of course, but it may be considerably more. For example, if a contractor from Texas (which has no income tax) goes to work in a state with an income tax, he is going to find a sizable chunk of his income disappearing—perhaps 7 percent or more. Forty dollars an hour in Texas is worth much more than $40 an hour in New York.

3. HOW MUCH ARE THE STATE AND LOCAL SALES TAXES?

This is an indication of how much more prices will be. The higher the sales tax, the more you will pay for living expenses while you are away.

4. HOW MUCH IS ACCOMMODATION?

A contractor from Dallas (where you can rent a nice one-bedroom apartment for $300 or $400 a month) would be shocked by California's accommodations costs. You may want to add a dollar or two to your hourly rate to pay the higher rent.

You also may want to go to a library or bookshop and look at Rand McNally's *Places Rated Almanac*. It will give you a good idea of relative costs throughout the United States; it also lists state income tax rates. These questions about living costs and taxes are essential if you are road-shopping. **Don't leave home until you know the answers!** If you live in Gulfport, Mississippi, and think that the agency's offer of $28 an hour is excellent, just wait till you get to San Jose, California, and see what the money buys!

5. WILL THE AGENCY PAY MOVING EXPENSES?

Some agencies will pay part of the expense of travelling to the contract, so ask how much they will pay.

Now that you've asked these questions, you should have a good idea of the contract the agency is offering. If you get past this stage, and interview with the client, you should remember to ask some of these questions again. Sometimes the client will not interview you—you may be offered the contract before you ever see or speak to the company. But before you sign the contract, you should try to talk to the person you will be working for. Ask the agency for the person's name and phone number and call. You should ask the client how long they expect the contract to last, exactly what sort of work you will be doing, and if the client expects or allows any overtime. And if you have any special concerns, as discussed in number 14 above, ask about those too.

You must ask these questions for several reasons. First, the agent you are working with may not understand your business. When the client gave the agency a list of requirements the agent may have misinterpreted the type of work you will be doing, and accidentally given you the wrong impression. Second, the agent may make a mistake. They work with dozens of clients simultaneously, and an agent may simply get two mixed up and tell you a contract is a year when it is actually three months. And, finally, some agents are simply unethical. They may lie, or just exaggerate—if a client said a contract probably will be three months but may last six, the unethical agent might tell you it is a six-month contract.

When you discuss a job with an agency, don't expect to get the contract—you have a job lead, not a job. When an agency tells you that you probably have the job, or even that you have the job, don't believe it **until you have signed the contract**! You should continue talking to other agencies. You can tell the other agencies that you probably have another job lined up, but you are still willing to consider other positions until you have signed the contract. This will put you in a strong bargaining position with those agencies. If you have been in any type of sales work you will understand this. Until they have signed the contract (and, in some types of sales, until the commission check arrives), most experienced salespeople do not assume they have a sale, regardless of what the client says. Too much can go wrong between the client saying yes and the signing of the contract. The client's boss may say no, the purchase order may get bogged down in red tape, the client could even resign or get laid off—honestly, I've had it happen! Worse, when the agency tells you that you have the job, they may be lying.

Most agencies want you to think they have the contract, and that all they need to do is find the right person. More commonly, though, an agency is competing against several other agencies, or even against the client. In other words, if the agency can find a good person at the right price, the client may buy; but the client is also looking for contractors from other agencies, and even looking for independent contractors. Whatever the situation, it is rare for a company to "buy" contractors from just one agency, so when an agency tells you that the local IBM office or Boeing plant only deals with them, you can usually discount it as nonsense. If you do run into this type of misrepresentation, I wouldn't recommend that you refuse to do business with the agency. Just keep quiet and humor them. Unfortunately, so many of the agencies exaggerate their abilities that if you refuse to do business with the ones you catch in a lie, you'll be left with a small number of agencies— that you haven't yet caught! No, your best defense is simple: Know your market, know how to deal with the agencies, and don't believe anything until you see it on paper (or better still, until your new client shows you where your desk is.)

I'm going to finish this chapter with a reprint from an article originally published in *PD News*, titled "What Is a Fair Rate?" You will hear contractors complaining about the agencies, claiming they got "ripped off," or that the agency doesn't pay enough. Certainly many agencies will try to get you as cheap as possible and sell you for as much as they can, but it's up to you to market yourself properly.

What Is a "Fair" Rate?

Is there any such thing, in a free market economy, as a "fair" rate? After all, isn't "what the market will bear" the closest we can get to a definition of "fair"?

Of course I'm talking about job-shop hourly rates here, and there appear to be two basic positions on the subject. On one side, many people believe there is such a thing, that job shops should not offer less than the "fair" rate for a job, although no one seems to be able to tell me what that rate is. Most job-shoppers seem to fall into this category. They don't know how to describe a fair rate, but they know it when they see it, and they are quick to complain if they are not getting it.

Many on the other side of the fence, the people running the agencies, have a different view: If a job-shopper is willing to take a job at the stated rate, it must be fair. They've got a point, too. After all, no one is forcing you to take the job.

There's another party in this relationship, though: the client. I would bet that most clients assume the bulk of the money they pay an agency goes to the contractor. Perhaps that is a good rule of thumb: "The agency gets a cut of what the contractor makes, not the other way around."

Of course that doesn't always happen. Let me tell you about a situation I observed in Dallas recently. This took place in the offices of a well-known telecommunications company. I can't tell you who, but you would recognize the name if you heard it. One agency had managed to place three or four technical writers in a documentation department on long contracts.

The rates were good for writers in Dallas—one contractor was earning the agency $43 per hour. But all the "contractors" were pseudo-salaried. The agency had found writers inexperienced in the world of contracting and persuaded them to take the kind of hybrid salary/hourly rate I described in an earlier article. The result was that the agency was getting quite a deal. The writer they billed for $43 an hour actually cost them about $18, including benefits—quite a nice spread.

Now, as we all know, contractors like to gossip, and this particular department had several writers making a lot more than $18 an hour. It also had trouble keeping information confidential, so it wasn't long before the lowest-paid contractors knew not only what they could make on a good contract, but how much the agency was billing for them. Not a healthy situation.

To make a long story short, people started looking elsewhere for work, for higher rates. Before long the client learned what was going on and, not surprisingly, was angry to learn that the bulk of what the agency was charging was paying for the agency's overhead and the salesman's Corvette.

Most clients are not unreasonable people. They know everyone has to make a profit, that agencies are not charities, but they also assume that most of the contract money is going to pay for experienced, skilled people; after all, if they can get people at less than half the price, why the hell pay $43 an hour?

Perhaps it would be poetic justice if the agency had been hurt by what it did, but I'm not sure it was. The client forced the agency to pay the contractors more, but it had been making over $20 an hour per contractor for almost a year already.

Sometimes agencies do get hurt, though. A few weeks ago a friend of mine came to the end of her contract. "No problem," said her agency, "we've got a contract for you," and promptly set her up with a job scheduled to begin the day after the old one ended. Just one problem: a rate so low that she would be better off taking permanent employment. Now, the agency may have miscalculated and bid too low, very low. However, I learned from other sources that the client company was willing to pay quite a high hourly rate, enough for the agency to pay my friend $7 or $8 an hour more, and still make $7 an hour themselves (yes, after payroll taxes).

The client offered my friend the job and she accepted it. But she found a better contract, and jumped ship the day before the new one started. The agency scrambled around for a substitute, but before they could find one the job went to another agency (paying considerably more money, by the way). The second agency may be making a small cut, but a small cut of a lot is a lot more than a large cut of nothing.

I don't much like the agencies making such large markups, but sometimes I think the "free market" theorists are right. I've seen contractors accept bad rates not because they had to, not because there was no food on the table and they had a sick child at home, but because they were not prepared to make the effort to find a good rate. Many contractors don't know what rate their skills and experience can bring—a situation they can easily remedy by talking to other contractors in their field. Often contractors know how much a good rate is, though, but still settle for less. Why?

The problem is they don't market themselves well enough. Too few job-shoppers realize they have to sell themselves to not one but two groups of people: the client and the agency. Too many job-shoppers believe that marketing is what the agencies are for, so they can just sit back and wait while the agencies deliver contracts to them. Yes, you can work like that, but if you work with just one agency, you had better have an excellent relationship with a very good one, or just resign yourself to paying a lot of money for their services.

If you want to do well as a contractor, you have to be a salesman. You might believe that only independent contractors need sales skills, but if you are getting an "unfair" rate, maybe you had better examine your own ability to sell yourself before you complain about the ethics of your agency.

11
Unethical Agencies

So far I've explained how the agencies can help you. Sure, you need to know how to negotiate, how to make the best deal, but that's business. I am now going to tell you something that will upset many people, perhaps even my friends in the technical service business. I'm going to explain why using technical service agencies is often not in the best interests of the client companies.

There are two types of problems with technical service agencies. One type is the unethical behavior of many agencies, and the other is the inherent weakness in the whole system of hiring people from an agency.

Now, I should point out that the problems I discuss in this chapter hurt the client company, and only occasionally hurt the contractor. But it is always a good idea to understand how the business in which you are working operates, and there are specific reasons that you should know about these problems.

For example, if you work through an agency, that agency's unethical behavior can tarnish your own reputation, however good you may be. If you are an independent contractor, you will find many companies unwilling to hire you unless you go through an agency. The information in this chapter may provide enough ammunition to persuade your potential client that the agencies are not such a good deal. And if you are ever in the position to employ contractors, you will know why you should hire independents. (I discuss how you can do so legally in chapter 21.)

Before I start, I want to apologize to any friends or associates whom I may offend. Some agencies work in the most ethical manner possible, but many don't. If what I say offends you, please assume I'm talking about someone else's agency!

The first thing a client should remember is that the agency is selling something. When an employee of a corporation has to hire another person, he or she usually wants to hire the best person for the job (the occasional case of nepotism aside). But if that employee decides to allow technical service agencies to compete for the business, another factor has entered the picture—the profit motive. Although the company employee does not profit from hiring someone, the agencies do. They don't want the client to hire the best person available, they want the client to hire one of *their* people. (If that person happens to be the best available, that is fine, but if he isn't—so what?)

Think of buying a car as an analogy. You wouldn't go to a car dealer and say, "I want to buy a car, find me the best one." You would find that the Ford dealer wants you to buy a Ford, and the Isuzu dealer claims that the best car is an Isuzu! That's why you educate yourself before buying a car. You may read car magazines, look at several models, sit in them, even test-drive them. You may research the dealer price of the car, and ask other owners how they like the vehicle.

If you want to hire a contractor you also should gather much of the information yourself, and not rely on the biased opinions of an agency that stands to make tens or even hundreds of thousands of dollars.

Here are eleven reasons companies should avoid the agencies.
1. Most agencies don't check references.
2. Some agencies do check references—but it may not help the client.
3. Competition among agencies hurts the clients.
4. Many good contractors will not work with agencies.
5. Many good contractors charge more than agencies will pay.

6. The agency may keep 60 to 70 percent of the pay.
7. The agency may encourage the client to take a less qualified person.
8. The agency may stop the client from rehiring the contractor.
9. The agency may stop the client from offering the contractor a permanent position.
10. Some clients don't want independents, so they hire through agencies—and unknowingly get independents.
11. Clients can hire independent contractors, and save 20 to 50 percent.

1. MOST AGENCIES DON'T CHECK REFERENCES

All agencies will deny this, but it's a fact. Some agencies do check the references of potential contractors, but most do not. I know this to be true from my own experience, and from talking with other contractors and agents. I have been sent on interviews (and been accepted for jobs) when the agency has not checked my references. My references are important to me, because I get good ones. So I always offer my references when I go for an interview. "That's okay," one client told me, "I let the agency handle all that." Of course the agency didn't.

This is how the agencies operate. They discover that a company needs a computer programmer, for example. They check their résumé files, and call all the people who appear to have the necessary qualifications and experience, asking if they may submit the résumé to the client. They also may ask for the names of other people who could do the job, and then call those people. (Some agencies will submit a résumé without even asking the contractor first.) The agency then submits as many résumés as possible, and arranges interviews. The agency does all this in a very short time, possibly a few hours even, because it knows that the competition is hard on its heels.

Let's imagine that the client wants to hire one of the programmers. Remember, the agency has not had time to check references. It is being offered a profit of $1,000 a month, maybe even $2,000 or $3,000. The agency doesn't have to work for this money—they have done most of the work already! So, what does the agency do? Check references and discover that the programmer isn't worth minimum wage? Risk throwing away thousands, even tens of thousands of dollars? Absolutely not! They accept the contract and hope that the programmer meets at least minimum requirements.

2. SOME AGENCIES DO CHECK REFERENCES—BUT IT MAY NOT HELP THE CLIENT

Some agencies do check references. You must remember, however, that agencies are usually competing against each other. As I've already mentioned, the agency doesn't want you to hire the best contractor; they want you to hire *their* contractor! So long as the contractor isn't a mass murderer, and turns up for work at least some of the time, the agency may still submit his résumé. (I know of an agency that—knowing what the contractor was like—submitted a contractor who only turned up at work 50 percent of the time. They got the contract, too, at least for a few weeks.)

Because references carry so little weight in the contractor-search process, undue importance is given to the résumé and the interview. Anyone can have a good résumé. If few companies check references, the contractor knows there isn't much chance of being caught in a lie. And many people *do* lie—or merely exaggerate—because they know they can get away with it. Clients looking for contractors generally use short interviews, perhaps believing that the agency is screening the people anyway. And don't we all know incompetent people who do well in interviews? Although John Gardner's Law (that 87 percent of the people in all professions are incompetent) may be an exaggeration, there *are* incompetent people in the workplace, and it is not great references that are getting them the jobs!

3. COMPETITION AMONG AGENCIES HURTS THE CLIENTS

In a free-market economy, a statement like that is almost heresy. But competition means the agencies have to cut corners. If a company is looking for a contractor without the assistance of an agency, the company will continue looking until they find the best person. The agencies, however, don't have enough time. They have to find someone they can sell to the company before a rival agency does. They have to submit as many people as possible as quickly as possible. The agencies provide the company with information about the contractors. They give the client a résumé, which they may have modified to fit the client's requirements. They tell the client that the contractor is good, perhaps even say that he has good references, and tell the contractor what the client is looking for, so the contractor knows how to respond in the interview. But when several agencies are submitting people, they can't all have the best person for the job. If the client company gets all its information from the agencies, it must decide based on interviews and the sales abilities of the different agencies. The agency that has both a skilled salesperson and a contractor who does well in interviews has a distinct advantage.

4. MANY GOOD CONTRACTORS WILL NOT WORK WITH AGENCIES

Many contractors can find work, at higher rates, without the help of the agencies. For example, in the Dallas area good technical writers may be able to get $35 to $45 an hour with an independent contract. The same writers working through an agency will get $28 to $32 an hour. So experienced writers who have built a strong network avoid the agencies. They only use them if they have to, and even then probably will keep listening for a better deal. Not all of these people are good at their jobs, of course—many will have excellent sales and networking skills instead—but generally these people are more experienced and skilled than the average contractor. So a company that uses only agencies (and will not accept independent contractors) loses a pool of talented people.

5. MANY GOOD CONTRACTORS CHARGE MORE THAN THE AGENCIES WILL PAY

Many contractors can only work with a few agencies because they demand a higher rate than most agencies will pay. Many agencies, especially some larger ones, do not pay very good rates. That doesn't necessarily mean that the client gets a low rate; it usually means that the agency has a very high overhead or makes a large profit. An agency I once worked for had one person in overhead (salespeople, secretaries, administrators, and so on) for every four or five contractors. After paying for those people, for their office space, desks, computers, and so on, there wasn't much left for the contractors. Such agencies pay lower rates than agencies with low overhead. But some agencies with low overhead *also* pay low rates—because they want to make a higher profit. An agency that bills $43 an hour for a contractor may keep $6 or it may keep $25.

Now, the client doesn't know how much profit the agency is making, or how much overhead it has, and usually doesn't know how much money the contractor receives. So the client doesn't know if the agency pays well or not. Of course, many clients don't care. They should, though, because the lower-paying agencies have fewer qualified people to choose from.

6. THE AGENCY MAY KEEP 60 TO 70 PERCENT OF THE PAY

Although clients often presume that contractors make good money, that is not always the case. One company I worked for, for example, paid my agency $37 per hour for my services, of which I received $28 per hour. At the same time they paid another agency $43 an hour for a less experienced writer, of which the writer received about $16. How, you may ask, does this hurt the client? Well, when the contractors learn how much money the agency is making and how much other contractors earn, they get disgruntled. The quality of their work declines, and they may look for a contract elsewhere, leaving the client in the middle of a project. At the very least, the contractor will waste the client's time and money in "bitch sessions" and on the phone to other agencies.

7. THE AGENCY MAY ENCOURAGE THE CLIENT TO TAKE A LESS QUALIFIED PERSON

If an agency has, for example, two contractors whom the client is equally interested in, and one of the contractors is demanding significantly more money, what does the agency do? If the client picks the one who costs the agency less, the agency will make a greater profit. So even if the more expensive contractor is also significantly more experienced or more skilled, the agency may encourage the client to take the cheaper one, though the client will pay the same.

8. THE AGENCY MAY STOP THE CLIENT FROM REHIRING THE CONTRACTOR

Companies may want a contractor to return to do more work later, but the agency may not allow this. First, if the contractor is on another job through the agency, the agency probably will not want to risk losing a contract by asking the other client to release the contractor. Second, if the contractor is no longer working through the agency, they will usually allow him to return to the client if only *they* get the contract. Most agencies have a clause in their contracts stating that the contractor cannot return to work for the client within a stated time without the permission of the agency; the contractor may have to wait a year or eighteen months before returning. If the company has to rehire the contractor through the agency, the contractor may turn down the contract if the agency doesn't offer enough money. This is especially likely to happen if the contractor has since begun working as an independent.

9. THE AGENCY MAY STOP THE CLIENT FROM OFFERING THE CONTRACTOR A PERMANENT POSITION

The contractual clause mentioned in 8, above, often stops contractors from taking permanent positions with a client. The contract usually has a clause that allows the client to "buy" the contractor, so long as the agency agrees.

If the company wants to convert a temporary contract into a full-time position, and the contractor accepts, the agency may not allow it. Or if the agency does allow it, the client may have to pay a fee—30 to 40 percent of the first year's salary. That may translate into $10,000 to $20,000, often enough to kill an offer. Some agencies simply will not allow clients to buy their contractors, believing it sets a bad precedent. Such contractual clauses may be unenforceable in some states; even so, they may still be enough to frighten off clients who are unsure of their legal position.

10. SOME CLIENTS DON'T WANT INDEPENDENTS, SO THEY HIRE THROUGH AGENCIES—AND UNKNOWINGLY GET INDEPENDENTS

Some clients avoid hiring independents in order to avoid the legal problems I describe in chapters 16 and 21. So instead, they hire through agencies. While most agencies hire contractors and pay them as employees, some do not, especially small "Mom and Pop" agencies. The agency takes the money from the client and pays the contractor "on a 1099." This means the agency does not have to pay payroll tax or get involved in complicated paperwork. But this setup is usually illegal, and leaves the client company—thinking it is safe—at risk. Contrary to popular opinion, the fact that the *agency* hires and fires the contractors doesn't mean the problems are the agency's alone. The IRS and the Labor Department recognize something called *joint employment*.

For example, Labor Department publication WH 1297, *Employment Relationship Under the Fair Labor Standards Act*, says this:

> Employees of a temporary help company working on assignments in various establishments are considered jointly employed by the temporary help company and the establishment in which they are employed. In such a situation each individual company where the employee is assigned is jointly responsible with the temporary help company for compliance with the minimum wage requirements of the Act during the time the employee is in a particular establishment. The temporary help company would be considered responsible for the payment of proper overtime compensation to the employee . . . Of course, if the employee worked in excess of forty hours in any work-week for any one establishment, that employer would be jointly responsible for the proper payment of overtime as well as the proper minimum wage.

As you will see in chapter 16, any agency paying a contractor on a 1099 may be breaking the Fair Labor Standards Act, because the agency is not paying time-and-a-half overtime rates. That means the client company, acting in good faith and assuming it has no responsibility, may be liable for the unpaid overtime and a penalty of an equal amount.

The client may actually be at *greater* risk using an agency than hiring an independent contractor. With the independent contractor, the client may be able to use the Section 530 "Safe Harbor" described in chapter 21, but once the client has hired an agency, the safe harbor disappears.

11. CLIENTS CAN HIRE INDEPENDENT CONTRACTORS, AND SAVE 20 TO 50 PERCENT

The best reason for a company to avoid agencies is that it just doesn't need them. By spending some time looking for contractors and checking references, a company can get better people for less money. How does the client do this? The same way the agencies do. A few classified ads in the local paper, a call to the local professional association job banks, and a call to potential contractors already known to the client will soon get the word onto the grapevine. And because there is no agency taking a cut, many of these people are cheaper than less qualified people hired from an agency. A writer who charges $35 an hour in an independent contract could cost the client $43 or more if an agency is taking a cut.

The cost of looking for a writer yourself is minimal, the savings enormous, and the time spent is very little when compared to the problems that arise when a company staffs a critical project with unsuitable contractors.

There are ways for companies to work with agencies and avoid most of these problems. They can tell the agency how much the contractor should get paid, and negotiate how much the agency will get on top of the contractor's hourly rate. They can demand to see written records of reference checks, and even check one or two of the references themselves. They can interview thoroughly, perhaps having several employees talk to the interviewee. And they can demand nonrestrictive contracts—contracts that allow them to hire the contractor again later or offer the contractor permanent employment, possibly after several months' working through the agency. They can ensure that the agency is not using independent contractors. And, perhaps most important, the client should simply remember that the agency is trying to sell something.

These are ways to improve the agencies' performance, but I believe the client is generally in a better position if it hires the contractor directly, without a middleman. The agencies will tell you they take the drudgery of searching for contractors away from the client; that they handle the payroll, and keep the IRS off their client's back; that it is difficult to find good people. All this is true, but the client must decide whether it is worth the additional cost—$5, $10, or $15 for every hour that the contractor works—and the potential problems.

12
The Interview

Once an agency has a client interested in you, they will arrange for an interview. This doesn't always happen—sometimes an overly trusting client will allow the agency to staff a particular project, and take the agency's word that all the people are qualified.

Normally, though, the client will want to talk to you first. There isn't a great difference between an interview for a contract job and an interview for a permanent position. The client wants to speak to you and get a general impression of what you are like and whether you will be able to do the job. The things that would encourage a client to hire you full-time are the things that will encourage him to hire you for a contract.

Before going on the interview, talk to the agency and get as much detail as possible. When the agency first talks with a client they ask for information about the type of person that the client is looking for: the profession, of course, but also such things as the type of projects the contractor should have worked on and the type of equipment he must have used. For example, a client looking for a technical writer may specify that the writer must have written hardware documentation and knows how to use desktop publishing programs. The specifications may be even more specific than that: the contractor must have worked on central office telecommunication switches, and know how to use Ventura Publisher and Microsoft Word. The client also may specify the amount of experience required (entry-level, three to four years, and so on).

All this information can be used to your advantage. Find out everything the client told the agency, and remember to stress the experiences you have had that most closely fit the ideal that the client is looking for. For example, you may not have worked on exactly the type of equipment the client produces, but you may have worked on a similar product, or a product that shares certain features with the client's. You may not have worked with the same tools or computer software that the client has, but you may have worked with similar ones. Even if you don't fit the bill exactly, you may still be able to get the contract; if you have enough experiences that are similar to those on the client's checklist, and if you appear to the client to be the sort of person who picks up information quickly, you may get the contract anyway.

It is important to appear confident in these interviews. As one contractor wrote to me, in response to an article in *PD News*, "The shopper must be totally confident in his ability to handle a project professionally and with little or no guidance." You must use the interview to convince the client that you are able and confident of your abilities, and don't need "hand-holding." Even if you are entry-level—as I've pointed out elsewhere in this book, even newcomers to a profession may be able to work on contract—you should still project an image of competence and ability. The client wants someone who can be trained quickly; someone who needs to be told something only once, not three or four times.

You can create a good impression by asking a lot of questions; don't overdo it, of course, but most clients are happy to tell you about the project. Ask questions such as these: What is the project? How long do you expect it to last? What skills are you looking for? What sort of tools and techniques are you using? How big will the team be? Has the company done similar work before, or is this a new venture? The more questions you ask, the more will occur to you, and the interview will quickly turn into a conversation.

You will find that many clients don't know how to conduct an interview. They usually leave that up to their human resources department, because although that department hires permanent employees, most company departments hire contractors directly, bypassing human resources. That leaves the manager not knowing what to ask, apart from a few obvious questions. If you "lead" the interview you will give an impression of being enthusiastic and intelligent. When the client mentions aspects of the project that are similar to other work you have done, tell him about that work, as if in passing, and then ask another question about the job, perhaps a question that will play on the similarities between the client's project and the other work you have done.

Another advantage of this approach is that you quickly get on friendly terms with the interviewer. Rather than continuing an interrogation that makes both you and the interrogator uncomfortable, asking plenty of questions and commenting on the answers turns the interview into a conversation, even a chat. And, as any salesperson should know, many purchasing decisions are made for personal reasons—whether you are selling computers, encyclopedias, or your time. The client doesn't think "I like this person, I think I will buy from him," but if the client feels comfortable and relaxed he is more likely to purchase from you—after all, he has to buy from someone, why not someone he likes?

This principle is even more important in contracting. If you are selling encyclopedias the client may purchase even if he doesn't like you (after all, he probably will never see you again), but if a client is thinking of buying your time, he's not going to hire you if he dislikes you, because he will have to spend every workday for months with you. Personal relations are critical in sales, so getting on friendly terms with your client as quickly as possible is essential.

There is another important reason to be on good terms with the client. Even if you don't get this job (as much as you charm the client, you simply may not have the experience required), the client can become an (unknowing) member of your network. Keep the client's business card, and call back when you are looking for work. Even if the client doesn't need your skills now, he may in the future—after all, he interviewed you, so you must at least be in the ball park. It would be unethical to bypass the agency and go to work for the client a week or two later, but there is no reason you can't keep in touch and go to work for him in six months, cutting out the agency to get a larger slice of the pie. Even if you never work for this client, he may still be a source of job leads.

Asking questions sometimes puts you in the role of adviser, even consultant. I have had clients ask me how I would handle a particular project, not as a way of testing me but because they genuinely were not sure how to get the job done. As a writer, for example, I may be interviewed by a small company that needs software manuals but has no writers of its own—that is why it is looking for contractors, of course. It doesn't know how to put these books together, which makes it very difficult for the company to interview contractors—how do they know if you can do the job if they don't even know what the job is? Acting confidently puts the client at ease; like a small child that is lost, they want someone to take over, to tell them what to do. If you can come in and act "quietly confident," showing that you know how you would do the job and are sure you can get it done, the client will relax and feel good about hiring you. Of course, you must take care not to be overconfident or arrogant or the client may wonder if you are all talk and no action. I'm certain, though, that a contractor who meekly sits back and waits for the client to take a commanding role won't get anywhere—clients are looking for someone to take the job off their hands, not someone who needs constant attention.

Naturally, many companies are looking for the exact opposite: They want a cog for their machine, not a power source. The client may simply want to see that you have all the qualifications on the checklist, and not worry too much about qualities such as initiative, confidence, and independence. This sort of interview might just be a way to look you over, to ensure you have no obvious personality disorders that would disturb the atmosphere (you don't appear to be alcoholic, psychopathic, or schizophrenic), and to make sure that what is on your résumé is correct. The interview could just be a formality; don't swear or spit, and you're in. This sort of interview is most likely for large projects, such as those undertaken by weapons manufacturers, when a company is hiring dozens, even hundreds, of contractors at once.

    Remember to ask some of the questions you have already asked the agency (the ones listed in chapter 10). Make sure you are getting the full story from the agent; the agent may not know exactly what is going on at the client's office, or may intentionally paint a rosier picture than the facts warrant. For example, the agent may say it's going to be a certain six-month contract, whereas the client may say it's three months with a possibility of an extension of two or three. An agency may imply that you can work as much overtime as you like, when the client will allow only three or four hours a week. Make sure that the impression you have of the contract is the correct one.

13
Contracts

If you are working through an agency you don't need to draw up your own contract, because the agency will have its own. Read it carefully, of course, and make sure you know what it contains, in particular how much you will be paid and how often. You should be paid at least every two weeks, and some agencies even pay every week. Also, check to see what it says about overtime, whether it states you will receive time-and-a-half for all hours over forty. Many, probably most, agency contracts contain a restrictive covenant, which seems to create more discussions among contractors than any other part of the contract. For example, one agency I worked for made me sign a contract containing the following:

The Employee covenants and agrees that the Employee may not:

i) during the period of employment and for one year following the termination for any reason of the Employee's employment, solicit or sell, for his own account or for others, data processing professional services that are competitive with the services of the Employer, to any company for which the Employee or employees under his managerial control (where applicable) has solicited or performed any data processing services on behalf of the Employer during any part of the year immediately preceding the termination of his employment;

ii) during the period of employment and one year following the termination of the Employee's employment for any reason, work or render data processing professional services, for his own account or for others, for any customer or the Employer for which the Employee has performed any services (or for which other employees under Employee's managerial control performed services), during any part of the year immediately preceding the termination of his employment with Employer; provided, however, that this restriction on employment shall, in the case of multilocation customers, be limited to the location or locations of the customer in question in which the Employee performed service and offices within a fifty-mile radius of such location or locations where the Employee performed services; and

iii) during the period of employment and for eighteen months following the termination of the Employee's employment, either directly or indirectly, hire any employee of the Employer in any capacity whatsoever, for his own account or on behalf of any person or corporation other than the Employer, nor attempt to induce any employee of the Employer to leave the employ of the Employer to work for the Employee or any other person, firm, or corporation within the geographic area where the Employer now or hereafter does business.

What does all this mean? The agency is trying to ensure that you don't use your position with the agency to sell the same services to a client of the agency that you worked with while employed by the agency; that you don't go to work for a client of the agency that you worked for while employed by the agency; and that you don't try to steal the agency's other contractors. The second part of the clause is the most significant, because it limits you from going back to a client for up to two years after you leave it. For example, you work for client A until the end of May 1991. The contract ends and the agency finds you another contract, with client B. You work for client B until the middle of May 1992. You leave the agency at that point, but according to the contract you signed, you cannot go back to client A until the middle of May 1993.

Can an agency really enforce such a contract? Maybe, maybe not. Texas, for example, is a "right to work" state, which does not allow limitations on a person's right to work. That means, according to some contractors and attorneys, that such a clause cannot be legally upheld. One contractor friend of mine decided he wanted to continue with the client he was working for, but without going through the agency that originally got him the contract.

Of course he stood to gain financially, but what spurred him to action were what he believed were unethical dealings by the agency. His attorney told him to leave the employer for a couple of days and then sign a new contract. Of course he needed the cooperation of the client, but they agreed to write a new contract. The agency didn't challenge the deal, but whether this is because they didn't think the contract would stand up in court, or because they had a lot of business with the client and didn't want to risk losing it, I don't know.

Such deals are common. I know several writers who have broken these clauses and returned to a client. If you want to do so, remember that you can do it only if the client agrees; and the client, for a variety of reasons, may not do so. The client may believe it is unethical, and the client also may have signed an agreement not to "steal" you. Even if the agreement isn't legally valid, it may be enough to scare off a client. On occasion, though, there are very good reasons for breaking a contract with an agency, and you may be able to persuade the client to work with you, especially if the client likes your work. For example, if an agency is not paying you regularly, or has defaulted on some of its other obligations to you, such as providing health insurance, you may be able to convince the client to dump the agency. Before you do so, contact an attorney and discuss the possible repercussions.

Independent Contracts

If you work as an independent you will need some form of contract. (For the sake of simplicity I have included independents' contracts in this chapter, rather than put them elsewhere.)

If you are working independently, don't work without some form of written agreement. Samuel Goldwyn said that verbal contracts "are not worth the paper they are written on," and he was right. Putting your agreement on paper is not implying you don't trust the client, you are simply making sure that both parties know exactly what their obligations are. A written contract not only helps you avoid problems later, when memories of what was agreed upon begin to fade, but may be essential if you ever need to take legal action in order to recover what a client owes you.

Contracts don't need to be complicated, and don't need to begin with "Know all men by these presents" as one contract I've seen does, or contain terms such as "hereinafter set forth." They can be in clear English, and simply explain what the client wants you to do and what the client must do in return.

Many companies, in particular large companies, already have a boilerplate contract they use with contractors; read through it carefully to make sure there is nothing unreasonable before signing. Some companies insert clauses that protect company secrets, patents, and copyrights, and restrict the contractor from directly competing with the client (or using information gained while working for the client to help one of the client's competitors). These seem reasonable aims, but make sure the wording is not too restrictive; it would be unreasonable for the contract to stop you doing *any* work for one of the client's competitors, for example, or to claim the right to *any* patents you receive while working for the client (as many permanent employment contracts do).

If you are doing a set-fee project—a project in which you are paid an agreed-upon sum for the project, regardless of how long it takes—makes sure you get some money up front. For example, if it is a short project you might ask for a third before you start, a third halfway through, and a third when you finish. If it is a longer project you might ask for a monthly payment; for example, 10 percent a month for seven months and the rest when you finish.

Also, make sure the terms of payment are acceptable. Ensure that the contract provides a reasonable payment period (thirty days after you invoice the client, for example), and that both you and the client agree on how often you will be paid. You don't want to wait until the end of a three-month contract for your money, for instance—invoice every week or two instead. Also check that the contract doesn't contain any punitive conditions for unsatisfactory work. For example, make sure that the contract doesn't allow the client to stop payment on all unpaid invoices; if the client doesn't like your work he should end the contract quickly, not let it drag out for weeks and then refuse to pay. Some contracts even state that the contractor is liable for any expense the client may incur purchasing the services of other contractors in the event that you "default" on your contract.

If your client doesn't have a contract he wants to use, you might just type up a letter of agreement. All it needs to state is the length of the contract, how much you will be paid, how often you will be paid, and how the contract may be terminated. Write a simple letter, addressed to the person who is going to sign, such as the following (just replace COMPANY with the company's real name):

Date

Name
COMPANY
Address
City/State/Zip

Dear _____,

This letter confirms our contracting agreement, as we discussed when I visited your office yesterday.

We agreed that I would provide my services to COMPANY for the period between January 15 and June 15, 1991. I will write the User Manual for the new product COMPANY is developing. I will work approximately 40 hours a week, more if necessary and if COMPANY approves the extra hours.

For my work COMPANY will pay $40 an hour. I will invoice COMPANY every two weeks, and COMPANY agrees to pay the invoice within 30 days of receipt.

This agreement may be terminated by either party at any time upon five days' written notice. Upon termination, COMPANY will pay me for any hours I have worked for which I have not already invoiced COMPANY, plus any unpaid invoices.

If this is your understanding of our agreement, please sign and date one copy of this letter and return it to me.

I will look forward to working with COMPANY.

Sincerely,

CONTRACTOR

Accepted:

COMPANY

By: _____

Title: _____

Date: _____

If your client wants a more formal document, try something like this:

THIS AGREEMENT, made and entered into on this ____ day of _____ 19____ by and between COMPANY NAME (CLIENT) and CONTRACTOR NAME (CONTRACTOR), is to witness the following:

The parties agree as follows:

1. TERM
The term of the agreement shall be from the _____ day of _____,
19_____ until the _____ day of _____, 19_____.

2. SERVICES
CONTRACTOR agrees to prepare a User Manual for CLIENT'S new product. The manual shall conform to the guidelines set down by CLIENT and shall be completed to CLIENT'S satisfaction. CONTRACTOR shall work approximately 40 hours a week; if more hours are required to complete the work they must be approved by CLIENT.

3. CHARGES
CLIENT agrees to pay CONTRACTOR $40 per hour for each hour of work performed by CONTRACTOR. CONTRACTOR will invoice CLIENT every two weeks, and CLIENT will pay each invoice within 30 days of receipt.

4. TERMINATION
This agreement may be terminated at any time before the completion of the agreement upon five days' prior written notice. Upon termination of the agreement by either party, CLIENT shall pay CONTRACTOR for any hours worked for which CONTRACTOR has not already invoiced CLIENT, plus any unpaid invoices.

5. CONFIDENTIALITY
CONTRACTOR agrees not to disclose information relating to the products, trade secrets, methods of manufacture, or business or affairs of CLIENT that CONTRACTOR may acquire in connection with or as a result of any work performed under this agreement. CONTRACTOR will not, without prior written consent, publish, communicate, divulge, disclose, or use (except in the performance of the work specified in this agreement) any such information. This obligation of confidentiality shall not apply to information that:
1) becomes publicly known through no fault of CONTRACTOR
2) CLIENT approves for disclosure in a written document
3) CONTRACTOR rightfully possessed before disclosure by CLIENT, or
4) CONTRACTOR independently develops without use of confidential information.

IN WITNESS WHEREOF, each party has caused this agreement to be executed by its duly authorized representative on the date first mentioned above.

CLIENT warrants that it has full power and authority to enter into and perform this agreement, and that the person signing this agreement on CLIENT's behalf has been duly authorized and empowered to enter into this agreement. CLIENT and CONTRACTOR further acknowledge that they have read this agreement, understand this agreement, and agree to be bound by this agreement.

CONTRACTOR   CLIENT

_____   _____

Name _____   Name _____

Title _____   Title _____

Of course the client may ask you to add certain clauses. For example, in the termination clause the client may prefer this:

> This agreement may be terminated at any time before the completion upon five days' prior written notice, or immediately upon unsatisfactory performance.

This seems quite reasonable, although I omitted it because it isn't in your interest. You also may wish to omit the Confidentiality clause unless the client requests it. There are some other clauses you might want to consider; you could add a clause stating that *your* performance depends on being given adequate information and an up-to-date version of the product, for example. If you are documenting a product that is not yet finished you might specify a "freeze date," the date by which your client will give you a finished product (so you don't have to complete a book one week after the product is finished).

There are many more permutations available. Howard Shenson, in *The Complete Guide to Consulting Success*, lists the following types of contracts:

Fixed-price Contracts
   Firm Fixed-price
   Escalating Fixed-price
   Incentive Fixed-price
   Performance Fixed-price
   Fixed-price with Redetermination
Fixed Fee Plus Expenses
Daily Rate

Time and Materials
Cost Reimbursement
   Cost Contracts
   Cost-plus-fixed-fee (CPFF)
   Cost-plus-incentive-fee (CPIF)
   Cost-plus-award-fee (CPAF)
Retainer

It is rare for contractors and consultants to use these contracts, however, excepting the Fixed-fee and Daily Rate contracts (of course the Daily Rate is just a permutation of the Hourly Rate contract).

The Fixed-fee contract states that the contractor will deliver a finished product, or do the specified work, for a stated fee—it doesn't matter how long it takes you to do it, you still get the same money. Nevertheless, you should sign this sort of contract only if you have a good idea of how long it will take to do the job, or you might end up making very little money for your time. You may be able to renegotiate the contract once you realize you have underestimated the work (especially if the client also miscalculated), but don't count on it. Always avoid this type of contract if the project is not clear-cut and easily estimated. For example, if a client wanted me to document a product that was so far from completion that it was hard to estimate how much information would have to be put in the book, I would be wary of a fixed fee, because it would be difficult to estimate accurately the amount of work involved, and because my work would depend on the client's employees meeting their deadlines.

However, if you know exactly how long a job will take, a fixed fee may be a good way to go—clients like them, of course, because they know what their final cost will be, and they know there is no incentive for you to "drag the work out." Howard Shenson believes such contracts to be very profitable. According to a survey he conducted in 1978, consultants who charged fixed fees had profits that were 95 percent greater than those of their colleagues who charged daily rates. This type of contract is especially beneficial to those whose skills or experience allow them to complete a job in considerably less time than most other people in their profession. While a client may balk at an hourly rate twice the norm, he will be happy to pay a fixed fee that is close to his estimate of the costs, not knowing that you will finish the job in half the time.

Robert E. Kelley, in his book *Consulting, The Complete Guide to a Profitable Career*, lists twelve factors to be considered in writing a contract:

1. **Responsibility of each party:** What do you and the client agree to?
2. **Time agreements:** When will you do the work and when will the client complete his obligations (such as providing you with necessary information)?
3. **Financial arrangements:** How and when will the client pay you?
4. **Products or services to be delivered:** How will you deliver your work?
5. **Cooperation of client:** Will the client be required to provide cooperation in the form of information, access to employees or company facilities, or raw materials?
6. **Independent contractor status:** State that you are an independent contractor and do not have employee status or obligations.

7. **Advisory capacity:** If applicable, state that you will not be making decisions for the client, only giving advice.
8. **Client responsibility for review, implementation, and results:** Make sure that the client determines the quality and result of your work. Reviews can be especially useful, because they ensure you don't complete a project and *then* have the client tell you he doesn't like the result.
9. **Your potential work with competitors:** If you may be working with the client's competitors, you may need to state this or agree to limitations.
10. **Authority of client to contract for your services:** Some corporations restrict employees from entering into contracts, so the client should state that it is authorized to do so.
11. **Attorney's fee clause:** If you find it necessary to use an attorney to collect your fees, this establishes who will pay the attorney's fees.
12. **Limitations:** State any special limitations, such as limited liability after a certain date.

Kelley is addressing an audience of people who intend to consult, rather than contract. Most technical freelancers contract, and such contracts may be much simpler; you are providing your services by the hour, so the relationship is not as complicated as if you were promising to deliver a product for a fixed sum. Incidentally, although Kelley suggests you state in the contract that you are not an employee, this clause may be of limited use: If, for example, the IRS decides that you *are* an employee, the clause has no effect and the IRS can assess the client for back taxes. (See chapter 21 for more information.)

If you or your client decide that you need a more complicated contract, talk with an attorney, or at least have an attorney review the contract before signing.

For more information on contracts see Shenson's *Complete Guide to Consulting Success*, Jeffrey Lant's *The Consultant's Kit*, and Robert E. Kelley's *Consulting, The Complete Guide to a Profitable Career*. (See the bibliography in Appendix E.)

14
Buying Your Benefits

Working freelance means you will lose benefits that an employer usually provides. Don't forget to replace these benefits. Working without medical insurance, long-term disability insurance, or a pension plan may give you a lot of money to play with, but can be disastrous if you get sick or live long enough to retire.

Most people can replace these benefits, although someone with a stable job and serious medical problems should think hard before working freelance. Unless you find an agency with a good medical policy, you may be uninsurable.

Medical Policies

Medical insurance is probably the biggest problem for freelancers. It is getting more and more difficult to find a good medical policy, and once you have one you don't know for certain if you are really insured—until you get sick. Perhaps you have read about the problem recently, or heard about it on "60 Minutes" or National Public Radio; you may have a health insurance policy, but your insurance company may not honor it. Why? Well, they may go out of business before you get sick—or worse, soon after you get sick (then you can't get another policy). They may close down the "plan" your policy is on, which means they don't have to pay claims to you or any of the other sick people they had insured.

If you get seriously ill they may look for a way to dump you. The easiest way is to dig into your medical history to find something that you didn't put on your original policy application, and then accuse you of fraud. That's easier than you might imagine; there are often mistakes in medical records—illnesses that you have never had, for example. You can't possibly remember your entire medical history, and so much is open to interpretation anyway. The Kaiser Permanente application form, for instance, is three pages long, for instance, full of questions such as "Have you ever felt depressed?" "Do you drink alcohol?" and "Have you ever fainted?" Like many other such applications, this one also asks you if you have been treated for anything by a doctor in the last five years, and then gives you three lines to enter everything. "Five years" and "ever" are long periods to remember; most people would find it impossible to complete one of these forms accurately, and some insurance agencies take advantage of that, using your inability to provide all of the information as a reason to cancel your policy if you get seriously ill.

This is not just a problem for individuals, though; it's very difficult for small *companies* to get reasonable insurance these days. In fact, in some ways you are better off on your own policy than a small company's policy. Insurance companies are allowed to dump people only if they close down an entire *plan*. Generally, each organization is a plan, whether a business, professional organization, club, union, and so on. The bigger the plan the safer you are, because the risk is being spread further. For example, if a business has ten employees and one of the employees (or a dependent) gets cancer, the insurance company may cancel the plan, because the cost of treating the cancer will outweigh the profits that could be made from such a small group. Everyone in the company loses his insurance, and probably won't be able to get more insurance, unless the person with cancer agrees not to join the new policy. (There is talk of banning insurance companies from dropping plans in such cases, but at the moment they can do so, and often do.)

How do you replace your medical insurance, then? First, I recommend you buy *Winning the Insurance Game* (Ralph Nader and Wesley Smith), which gives a number of ways to protect yourself when you are shopping for insurance (for example, check the company's rating in *Best's Insurance Reports*). Read this book, and *then* look for medical insurance.

Of course, you should see if the agency you are going to work with has a medical plan. Many, if not most, agencies now have plans. If your agency doesn't, or you are working as an independent, you will have to find insurance elsewhere. Even if your agency *does* have a policy, though, you may want to consider getting your own policy, especially if it's a very small agency. In addition to the problems already discussed, moving from agency to agency could cause problems if you get sick. When you move to a new agency the new policy will probably have "preexisting condition" restrictions: The policy either will not cover, or will cover to a very limited degree only, conditions that you had before you signed on to the policy. Of course this is a problem that can occur to captive employees also, when they are fired or laid off, but the problem may be worse for contractors because they move between policies more often.

If you are a military veteran, of course, you may not want to bother with health insurance at all, if the military's medical services in your area are good. If you do need insurance when you go freelance, COBRA may help. The Consolidated Omnibus Budget Reconciliation Act forces insurance companies to continue medical coverage for employees who leave a company for any reason other than "gross misconduct." I don't know exactly what gross misconduct means, but if you are laid off or (in most cases) fired, you can continue your medical insurance for up to eighteen months. It will cost more than a company policy usually costs, because you will have to pay the full cost of the policy, paying both your share and your ex-employer's share (plus an extra 2 percent if the company wants to charge a service fee). You remain covered under COBRA until you are eligible to join another employer's plan, *unless* you have preexisting conditions, in which case you are still eligible for the full eighteen months.

This is an excellent way to keep medical coverage, and remember that if you work for an agency that has a medical policy you can extend that policy when you leave—so long as you joined the medical policy before you left, even if you were only on it for one day. As long as you are a legal employee of the agency (if the agency withholds taxes and reports your pay to the IRS on a Form W-2) and the agency has more than twenty employees on a typical day, you have the right to a COBRA extension. I extended one agency's policy for $260 per month, which included not only an excellent medical and dental policy for my family but disability insurance as well. (There is no federal law that states your employer has to continue disability insurance, but they may do so if it's all part of the same package.)

If you do decide to use your COBRA rights, you have sixty days from quitting to inform your ex-employer, and then another forty-five days to pay the first premium. The law requires all employers to provide information about COBRA to all employees leaving the company, so if you don't get this information, call the company and remind them.

If you are an independent contractor and you have problems getting medical insurance, you might consider getting a short-term contract every couple of years with an agency that has a good medical policy, just so that you can extend with COBRA. (You might even get a full-time job for a few weeks!)

There are some problems with COBRA. Your ex-employer may stop providing insurance altogether, in which case you are on your own. And your ex-employer may refuse to extend your insurance. This is illegal, but I'm told by insurance-law attorneys that it is quite common. If this happens, contact the Department of Labor immediately, and try to get them to help. If they won't, keep trying, because getting help depends on speaking to the right person at the right time. A local agency refused COBRA to a friend, and the Department of Labor forced it to comply. But when an agency refused to extend *my* insurance the Department of Labor told me to hire a lawyer.

How about your spouse's policy? If your spouse has a good company policy, check into being added to that policy; this is usually the cheapest way to get good medical insurance. Incidentally, if you live together unmarried, you may still be able to get on your mate's policy. If your state legally defines your relationship as a common-law marriage, your insurance company may be forced to accept you on the same conditions as a spouse. Many states have very simple requirements. Just living together for a few weeks may be enough. Of course, there could be reasons you wouldn't want to acknowledge such a union, but if you don't mind, check the definition of "spouse" with your insurance company and your state board of insurance.

Eventually you may find you have to search for an individual policy. Perhaps your COBRA policy runs out before you get a job with an agency that has a medical policy, or maybe you manage to find enough independent contracts and don't need to work through agencies anymore. So what next?

Finding an affordable medical policy is not easy, so give yourself plenty of time. Don't start looking a week before you need it. I suggest you start comparing policies at least two months before, preferably more. Insurance is a complicated business, and I don't have the space to explain all the options and possible problems, so read *Winning the Insurance Game*. Investing the time to read this book could save you much money and trouble in the future.

Probably the most affordable way to get a medical policy, if you can't get a policy from an employer, is to join some kind of "group." You might want to look at Workers Trust, an "association of democratically managed businesses and self-employed people." They have a medical policy that *Whole Earth* recommends. (See Workers Trust, Appendix F.)

A similar organization, Co-op America, also has a good policy. (See Co-op, Appendix F.) Both the Workers Trust and Co-op America policies are remarkably comprehensive, even covering services such as well-baby care, acupuncture, and homeopathy.

You can also join various organizations that have health insurance: the National Association for the Self-Employed, the Society for Technical Communication, the American Association of Retired Persons (AARP), professional associations, and so on. But there's something you must remember about these groups. In many cases these groups were begun *in order to sell insurance*. Even the AARP was—at one time—a "front" for Colonial Penn to sell insurance (see *Payment Refused*, by William Shernoff). These group policies are often no better than individual policies, so if an organization of which you are a member offers a group policy, don't assume it's the best deal. It may be a good policy, but check it out thoroughly (*Winning the Insurance Game* tells you how). In fact, when the insurance agent arrives at your house, make sure he's selling the right policy. When I met with an agent to look at the STC's policy I was offered the insurance company's standard health policy, with no modifications or discount.

Also, don't let the word *group* fool you, or the advertising which often implies that you can buy a policy that rivals that of a big company's policy—you can't. If you go to work for IBM (or most companies these days) you can join its medical policy without any medical examination or proof of insurability. To buy a "group" policy, on the other hand, you usually have to provide information about your medical history and may even need a medical examination. You can also have your policy cancelled if the insurance company finds mistakes or omissions in the information you give them. And many group policies have enormous gaps in coverage; they are so specific about what they *do* cover that an awful lot gets left out. If a plan has a long list of what it *does* pay, and itemizes what it does pay while you are in hospital, there are probably many things it *won't* pay, items that may not even appear in the "exceptions" section. Plans often have ridiculous limits (a maximum of $3,000 for surgery, for example, or $50 a day for the hospital room).

And be careful of plans that pay for hospital care but not for prescriptions outside hospital; you can spend a lot of money without ever seeing the inside of a hospital. (A friend told me of a policy he paid $340 a month for; it comprised six overlapping policies, didn't include medication outside the hospital or visits to the doctor, and ended up paying only 35 percent of his family's medical expenses.) On the other hand, buying a plan with "holes" is one way to reduce your premium, so if you can save $3,500 a year by buying a plan that will only pay $80,000 of a $100,000 medical claim it may be a logical move—so long as you understand exactly what it is covered and what is not.

Many plans are so confusing that it is difficult to figure out exactly what is covered and to what extent. They often have different levels and are made up of several separate plans that are combined, each covering something slightly different (and leaving out key items). This is no mistake. Insurance companies know how to make a plan look comprehensive, while carefully omitting important expenses.

All that having been said, there *are* some groups that have excellent health policies. There's a general rule of thumb, though. If *anyone* can join the group, the policy will not be too different from one you could buy outside the group—it may be discounted a little, perhaps. But if the group's membership is restricted, the policy is likely to be better. What sort of restrictions? In order to join some professional associations you have to prove you are employed in a particular industry, show you have worked in it for a certain number of years, be sponsored by one (or two or three) of the association's members, and send a copy of your official transcript or diploma proving you have a specific degree. The result is that the members are a very select group, often wealthy, middle-class men, so the association can offer it an excellent health policy. Its insurance company knows what to expect from the group, and knows that most of its members are in a low-risk group.

So spend a little time considering the groups you might be able to join. Ask your friends about professional organizations. The Society for Technical Communication offers a medical policy, but the STC has open membership, so the policy is not of the same standard as some professional associations' policies; in fact, in most states you will be offered a standard Mutual of Omaha health policy at a 5 percent discount. (In some states, such as Texas, Mutual has a special policy for the STC, which—a Mutual agent told me—it rarely sells because it is so expensive.)

You could also look for an HMO, a Health Maintenance Organization. You pay these organizations to keep you healthy, so you don't pay extra when you get sick. You pay a premium each month and get free medical care when you need it. The HMO owns the hospitals or has an arrangement with the hospitals to pay directly. However, the most common complaints about such organizations are that you must use one of the HMO's doctors, and you don't always get the care you need, because the people treating you have a vested interest in not spending too much time.

Once you find a medical policy you like, there are a few things you can do to keep your monthly premiums low. Buy a policy with a high deductible: $500, $1,000, or more. This reduces the premium considerably. If the policy comes with an added life insurance policy, consider removing it. One policy I bought had a $10,000 life policy for $6 a month; I cut it out because I already had enough life insurance, at a cost of only $1 a month for each $10,000 of coverage.

You also could cut costs by purchasing a policy that covers only major medical expenses; if you get the flu and go to a doctor, you pay, but if you get seriously ill and end up in intensive care, they pay. Omitting maternity insurance also cuts cost. Insuring against pregnancy can be very expensive, and often pays very little, or even nothing, for a couple of years. If you are planning to get pregnant within a year, you may pay more in extra premiums than you would get back in benefits—it's only after a year or two that most policies start paying any worthwhile amount. However, discuss this issue with the insurance company before you buy; some policies cover the *complications* of pregnancy, even if you don't buy the extra maternity coverage itself.

Another option you might consider is getting a policy with a very large deductible, often called a catastrophic care policy. Your deductible may be from $25,000 to $50,000, but your premiums are very low. A $35,000 deductible sounds like a lot, but if you are spending $5,000 to $8,000 a year in premiums for a family policy—and can buy a $35,000-deductible policy for about $500—you may want to consider "self-insuring" yourself for the first $35,000 (in fact many hospitals will write off a large portion of your expenses, effectively giving you a discount, so you may not even have to come up with the full $35,000). Over thirty or forty years you would save hundreds of thousands of dollars in premiums. So self-insuring for $35,000 per sickness may be a logical choice. Another advantage of such policies is that they are usually easy to join—you don't have to be examined or complete an extensive medical history form—so it may be hard for the insurance company to wriggle out of paying. Of course there are things to watch for with these plans. The deductible is usually per sickness, so you could end up paying $35,000 several times if you were really unlucky. (Some plans apply only one deductible, though, if more than one family member suffers in the same accident or from the same sickness.) Also, each sickness or accident may be covered for only a limited time (three years, for example). Some insurance agents can sell you these policies, but a lot of professional associations carry them as well.

Whatever you do about medical insurance, do something. The consequences of serious sickness *without* insurance can be severe, and financial ruin—or even death—is not uncommon (recent studies have shown that going to hospital without insurance doubles your chance of dying while you are there!). I'm stressing this in such strong terms because I know how easy it is to "let things slip," and I know several freelancers who don't have insurance. But it's just not worth the risk. Buy a book about the subject, and read a few magazine articles. Go to your library and check the periodicals index. You may find that one of the consumer magazines (such as *Consumer Digest*) has recently compared medical policies. If you are at all unsure about your insurability, investigate insurance long before you need it. (The bibliography in Appendix E lists several books on medical insurance.)

By the way, if you are an independent you can deduct 25 percent of your medical insurance premiums off your taxes on Form 1040. This deduction is not available to you if you are an employee. (This may not be extended after 1991, so check a current tax guide.)

Disability Insurance

Ever stopped to think about what you would do if you couldn't work? Disability insurance is usually more important than life insurance, because the chance of being *disabled* in the middle of your career is greater than the chance of dying. Most companies have some kind of disability policy, and many agencies now have such policies also, usually packaged with the medical policy. If your agency doesn't have one, though, you should start looking. A simple policy paying $2,500 a month after ninety days of disability could cost a 32-year-old man about $76 a month (I know because that is how much mine is). This policy also has a 4 to 6 percent cost of living adjustment to keep the policy in line with inflation, and the right to increase the monthly benefit (by increasing the premium) at any time, even if I become sick.

Of course the older you are, the shorter the waiting period, and the larger the monthly benefit, the greater will be the monthly premium. But there are also many "bells and whistles" that can quickly add up. You can purchase a Social Security supplement—a rider to the policy that pays extra benefits if you receive no money from Social Security—though the extra money usually lasts only two years. You also can buy a rider that pays a portion of the monthly benefit if you return to work part-time and a rider that pays the monthly overhead expenses of your business office while you are disabled. No doubt there are many other options, but buying all these can quickly double or triple the monthly premium.

You can get a disability policy from many insurance companies and groups, but try talking to SelectQuote first. They will give you a computerized list of low-cost disability policies at no charge. (See SelectQuote, Appendix F.) Look for a policy that defines *disability* as being unable to carry out the duties of your regular occupation (some policies are more strict than this, defining disability purely biologically, with no regard to the type of work you do—if you are fit enough to bag groceries, you are not disabled). Make sure you understand exactly how long the payments last if you are disabled. Some end when you retire, for example, and there are dozens of different permutations; mine would pay me for the rest of my life, unless I'm disabled by sickness after the age of sixty, in which case it pays until I'm sixty-five (if I'm disabled by *accident* after the age of sixty it still pays for the rest of my life). I've no idea who thinks up all these variations, but it must take some imagination.

Look for a noncancellable policy; so long as you pay the premium, the company can't dump you up to age sixty-five, and after sixty-five they can't cancel the policy so long as you are paying the premium and working. A policy with a *recurrent disability* clause may be worth considering, also. This means that even if you are not off work for consecutive days, so long as the disability is caused by the same accident or sickness and the days are not separated by more than six months, the insurance company will regard them as the same disability.

Disability policies are complicated things, and I can't make recommendations beyond these few simple points. Remember, however, that with any disability policy you will have a waiting period between becoming disabled and receiving money—another excellent reason to have money in the bank. Incidentally, although you can't use your disability insurance premiums as a tax deduction (except the portion, if any, that pays for business overhead insurance), any *benefit* you may receive is not taxable, so long as you paid the premium yourself.

Term Life Insurance

If you need a life insurance policy, you will generally need more than a typical company policy provides, because $10,000 won't go very far. Most employers' policies are nice little extras with little real value, benefits that don't need to be replaced directly, because you have probably bought a large policy independently. Of course, if your employer gave you a large policy—some insure your life for a couple of times your salary—you may not have bought your own. Before you go out and buy a policy, read Andrew Tobias's *The Only Investment Guide You'll Ever Need* or *What's Wrong with Your Life Insurance* by Norman Dacey. Both these books explain why you need a term life policy, not a whole life or cash value policy. And once you are ready to buy a term policy, don't go straight to an insurance agent.

First, check with your credit union to see if they work with a computerized listing service; these services take your information, feed it into a computer, and then give you a list of five or six policies and companies you can choose from. This is probably the best way to find low-cost life insurance. I used SelectQuote, mentioned earlier in this chapter. The service is free, and you can do everything over the phone (which is good if, like me, you hate hassling with insurance agents). You can even call and talk to the people at SelectQuote to get information and advice.

Tax-Free Savings Plans

There are several ways to replace your employer's savings plans. What do these tax-free plans do? They allow you to pay money into a pension fund, without first paying income tax on that money, and they allow you to keep all the income made by your pension fund, without paying income tax on it until you begin withdrawing money from the fund.

Your own pension plan has certain advantages over an employer's. Permanent employees rely to a great extent on the employer's contribution. That is part of the bargain: The employee receives less than the employer can afford to pay, but instead the employer puts some of the unpaid money into a pension plan. You may never get that money, though.

These days people seem to move around quickly. Sure, some people work with the same company for twenty-five or thirty years, but that is becoming less common. I know many people who stick with a company for two years and then get laid off, go to the next company and last eighteen months, and so on. But the employer contribution to a pension plan doesn't immediately become your property. Typically you must wait over three years to "own" twenty-five percent, four years to own fifty percent, and five years to own one hundred percent. An independent contractor probably can afford to save more money (because he receives more), and all the money placed in the account is his, with no waiting and wondering if he will ever get the money.

Also, many company pension plans are now going broke, and some retired employees are finding that their pensions are now worth only a few percent of what they expected—and what they were originally told the pension would be.

So, what can you do if you go freelance? First, you still have an IRA. Although recent tax law changes have placed some restrictions on IRAs, if you are not eligible for any other tax-free savings plan you can put the maximum amount allowable into an IRA: $2,000 for an individual, or $2,250 for a married couple when only one spouse is working.

If you are an employee of an agency, you may be able to use the agency's pension plan. Many agencies now have 401(k) plans or SEPs (I discuss these below). If your agency doesn't have any kind of pension plan, all you are left with is an IRA, so take advantage of it. Remember, though, just because you don't have a large tax-free savings plan doesn't mean you shouldn't save for your retirement. One of the major advantages of freelancing is that it allows you to earn more money; sure, your last permanent employer had a good tax-free savings plan, but did you make enough money to save any?

If you are an independent freelancer you can open two types of retirement plans: a Keogh or an SEP (Simplified Employee Pension plan). (If you work most of the year through an agency, but part of the year as an independent contractor, you can still set up one of these pension plans, but your savings are based on your income from self-employment only.) You can set up these plans through a bank, credit union, brokerage firm, or insurance company.

SIMPLIFIED EMPLOYEE PENSION PLANS (SEPS)

An SEP is an IRA set up to receive larger-than-normal payments. If you set up an SEP you can deduct up to 13.0435 percent of your income, or $30,000, whichever is less. SEPs are very easy to set up, and have the same rules as an IRA.

KEOGH PLANS

Keogh Plans, also known as HR-10 plans, allow you to deduct up to the lower of 13.0435 percent of your income or $30,000. You may set up a plan into which you can pay a fixed portion of your profits (a profit-sharing plan) or one into which you pay a fixed contribution (a money-purchase plan). You can set up both a profit-sharing and money-purchase plan if you wish. Keoghs have an advantage over SEPs in that when you retire and withdraw money from the Keogh you may be able to "income-average," reducing the amount of tax you will have to pay.

Okay, pension plans look complicated, but they seem worse than they really are because they have so many "ifs, ands, and buts" that won't normally concern you. Go to a bank (or better still, a credit union) and have them explain what you need to do. You will find it is reasonably straightforward. Or find a book on pensions and retirement planning, and invest a little time ensuring your financial security.

These are the most important company benefits that you must replace, and often the most difficult to deal with. Consider how you are going to replace these benefits some time before you need to do so, because they take time to investigate and set up. The health and life insurance are probably the most critical, but don't put off your pension plan. There is never a "good" time to start one, so it's easy to let it slip—but leave it too long and you may live to regret it.

Credit Unions

There is another benefit you may want to grab before you leave permanent employment. If your present employer has a credit union, and you are not a member, join before you become a contractor. Credit unions are generally in better financial health than banks and S&Ls, and can provide you with cheap checking accounts, high-interest savings accounts, and low-interest loans. Many even allow you to overdraw your account at little or no cost to you. One credit union I know lets you use a credit union Visa card to provide an overdraft on your checking account. When you overdraw, the money comes from the Visa, but without any cash-withdrawal fees. So long as you pay the balance by the payment-due date, there is no finance charge.

Everyone should have a credit union account, but freelancers will find the cheap loans and relaxed attitude about overdrafts especially useful. Incidentally, you may be able to join a credit union even if your employer doesn't have one, or if you don't have an employer. Look in the Yellow Pages, call all the credit unions, and ask them who can join. Some allow residents of a particular town to join, or people in a particular occupation (so if you live in Podunk, call the Podunk Community Credit Union first). You might also look at the National Association for the Self-Employed, which has a credit union for members (see Appendix C).

15
At Work

I know some would-be freelancers who are a little scared of working freelance; not because they can't handle money, not because they are scared of looking for contracts, but because they are not sure that they are "good enough" to freelance. Many people have a misconception about freelancing: They think you have to be experienced and highly skilled before you can be a contractor.

But this isn't so. I have met freelancers who were close to incompetent, but still kept themselves employed. Paradoxically, the technical service industry has many people who don't fit in well elsewhere, but can take advantage of unethical agencies. Because most agencies don't check references, many freelancers fabricate experience. A friend of mine, a manager of a technical writing department, once told me how she had interviewed a writer. It turned out that both had worked at the same company in the past, although not at the same time. She asked to see a writing sample, and he brought out a book he claimed to have written while with this previous employer. "I couldn't believe it," my friend told me. "It was my book!"

Another publications department manager told me, "I used to think contractors were experienced, skilled professionals. Now I realize that they are no better than anyone else, they just want to make more money." He wrote a memo to his boss explaining why he didn't like to use contractors. "They are not the skilled professionals they are advertised to be," he wrote. "I have grounds to suspect that they are more clever at marketing themselves and evading objective evaluation than producing good work."

Unfortunately he is quite correct. Don't misunderstand me, I'm not suggesting that incompetent people should become contractors (though obviously they can do so). What I am saying is that if you are reasonably hardworking and reasonably competent, there is no reason to think that you can't do well as a freelancer.

For example, I know one technical writer who has been a contractor since he left college two years ago. His first rate was $21 an hour, and the job lasted over a year (not a bad entry-level income, about $40,000). Another writer, in England, also began straight out of college, and has been a freelancer for fifteen years since. This lifestyle has allowed him to keep a boat moored in Greece, sail the Atlantic a couple of times, and spend a year sailing the Caribbean.

Often working as a contractor seems just the same as working as a captive. You keep the same hours, do the same work, follow the same instructions—but take home more money. You get invited to the company parties, picnics, and beer bashes. Some companies even include the contractors when they give their employees time off with pay ("Go to the product-release announcement this afternoon, and then go home early").

However, there is often a difference between the way your client treats you and the way he treats his employees. Remember that one reason the client hired you is that you are expendable. Your client can get rid of you quickly and easily, for any reason; he cancels the project, the project runs low on funds, or you don't do good work. Firing captive employees, on the other hand, is much more difficult, and we all know captives who seem to hold on to a job for years without appearing to do any work. You, on the other hand, have a contract that comes up for renewal every few months; your client has to justify the expense over and over again, so if he can't think of a good reason to keep you, he won't extend the contract.

Your clients are likely to expect a bit more from you (and perhaps be a little less lenient towards you) than they expect from their permanent employees. Also, don't underestimate envy—often you will be paid considerably more than your immediate superior, and this can cause some friction, perhaps even outright resentment.

Often clients are justifiably resentful. One manager referred to the "country club atmosphere" that pervaded his office full of contractors: people on the phone all day, people chatting with each other, swapping job leads, comparing rates. One contractor I worked with spent about three or four hours a day on the phone to friends, or just visiting people around the office.

Sometimes clients are partly to blame for these problems, though. Many offices have a very relaxed, laid-back atmosphere, in which very little work gets done. The client brings in a contractor or two, and expects them to do nothing but work. But it is very difficult for the freelancer not to get "infected" with the same lackadaisical attitude that everyone else has, especially when the contractor has to work with these people.

Some managers don't much like contractors, and only hire them because they have to. I know one department manager who doesn't think contractors should be allowed to have phones—he would like them to work every minute of every hour they bill. Unfortunately that is just not possible. Very few people can work eight hours straight without a break, and if anyone expects you to bill only for every minute you actually work, you had better increase the hourly rate at least 25 percent, because technical service contracts usually pay for normal work hours: an eight-hour day including a normal number of breaks.

You may find yourself without many of the comforts of the captive employee. Telephones seem especially difficult to get hold of for some reason (even if the manager has no particular objection to you having one). As a contractor you often end up at the bottom of the "to do" list. You also could find yourself moved around a bit, shuffled from one office to another as captives come and go. One company I worked for has a policy that denies contractors door-cards, those credit-card-size electronic "keys" that you hold up to a panel to unlock the door. Company policy stated that contractors would be let in by their manager each morning, a rule that my manager thought was as stupid as I did (he found me a card that had "fallen through the cracks").

Some contractors have the idea that they are consultants, that they are there to do the best job possible, and have a duty to tell the client the best way to get the job done. This isn't quite true. In the client's eyes you are a temporary employee, not a consultant. You are usually not expected to take over and run things your way. Sure, give your advice to the client, but if the client doesn't want to take your advice, don't get offended; it's the client's prerogative. Sometimes you just have to do things the wrong way, the inefficient way, because that is what the client wants. If you suggest another way a couple of times and the client isn't interested, get on with the job, the client's way. My attitude about per-hour jobs is that I get paid by the hour, not by the finished product. In some situations the client may know it's the wrong way to do a job, but other factors force him into doing the work that way—because the rest of the company does it that way, for example, or because his boss won't allow him to buy the equipment he needs to do the job properly.

This, incidentally, is the major difference between a contractor and a consultant. A contractor is paid by the hour, not by the job. The consultant, on the other hand, is paid to get a job done, and may even receive a lump sum for the job. In such a case the consultant *must* have the right to say how the job will be done, so it can be done quickly and efficiently.

So, how should you act at the company office? The rules are obvious, really.

1. Be prompt, and begin work when you are supposed to: Though many employees in your profession may be salaried, you are not, and the client tends to watch your promptness more than that of the captives. The client pays by the hour for your work, and is constantly reminded of this fact, and of the large sums involved, each time he signs an invoice. He can see his precious budget pouring out of his department, so each time he sees you goofing off, all he can think of is money going up in smoke. Of course captive employees cost a lot of money too, but that is more easily forgotten because the client doesn't have a constant reminder.

2. Avoid getting involved in a "country club" atmosphere: Don't spend too much time on the phone or talking with other employees or contractors. One company I worked for allowed employees to close their office doors. The result was small groups of employees and contractors "shooting the breeze" behind closed doors. Avoid these groups. The client may allow it at first, but will grow to resent it.

3. Do the best job you can: Sure, you can often get away with doing a mediocre job (thousands of contractors do so every day), but the better the job you do, the better your reputation will be. The client will be more likely to extend your contract, give you good references, and hire you again, and you will find it easier to get other contracts. Reputations travel.

4. Make sure your appearance is acceptable to the client: I hate ties, and avoid them if I can. One VP told me he didn't care what I looked like, so long as I did a good job (that is the sort of company I like to work for!). Unfortunately, many companies expect "business dress," which, for men, means wearing a piece of clothing that restricts the flow of blood to the brain. If that is what the company wants, that is what I do. Of course, even if you don't wear traditional business clothes, you should always be neat and well groomed. Acceptable dress varies among industries, professions, and companies. It even varies among different divisions of the same company. I once worked for a large Canadian telecommunications company that had a separate research division. People in the main company had to wear traditional business dress, while those in the research division wore jeans and sweaters, even though they worked in the same building.

So as much as I dislike business dress, I sometimes have to wear these clothes. I have come up with a good rule of thumb, though: Wear what you feel comfortable in, without upsetting your client. If your client doesn't care, there's no reason you should. (I know many people will disagree with that comment. I can hear them now: "You are a professional, so you should dress like one," they are saying. Well, yes, I suppose so, but I don't enjoy dressing like one. I would like a client to extend a contract or recommend me because I did a good job, rather than because I wore a grey suit.)

5. Don't drag your feet: I know contractors who expect to be waited on; they want the client to bring the work to them, not have to get up out of their seats and take some initiative themselves. They have the attitude "If the client wants me to do that, he's going to have to show me exactly how." But contractors earn a lot of money, and although the client rarely expects them to take over and work miracles, the client deserves, and often expects, a certain amount of energy and resourcefulness.

In fact, clients often expect you to have more initiative than their captives. After all, you are going to be there for a limited time, so the client doesn't want you to spend months "settling in" or "getting the feel of the project." You need to learn quickly: the client's word processing system, work procedures, specialized tools, report requirements, and so on. If you can quickly size up a situation and start producing something the client can see, you'll go a long way towards keeping the client happy.

Also, don't try to drag a job out, so that you can keep pulling in the paychecks. A good manager can see what you are doing, and after a while it becomes obvious to even the most trusting boss. It won't do your reputation any good, and if you market yourself properly it shouldn't even be necessary, because you can always find work elsewhere.

6. Remember that while you are working for the client you are still selling yourself: Even after you get a contract, you are always selling yourself. You are working to get the contract extended, the next contract, the good reference, the referrals to other managers, and so on. This doesn't mean you have to be hyped up all the time, acting like a caricature of a used car salesman, but it does mean you should try to get on with people. Try to avoid conflicts. Be helpful, friendly, and cooperative. Reputations are built not only on competence, but also on congeniality. Consider these two references: "He's good at his job, and he's friendly and easy to get on with" or "He's one of the best, but he's a real pain in the neck to work with." Which would you rather have? I'll bet that the first reference gets more jobs.

Successful freelancers are usually friendly people. It's very difficult for someone who doesn't get on with others to do well in freelancing, because he or she can't build a solid network. Each job you have is a stepping-stone, a way to find a useful contact, something to build on. If you leave behind a company full of people happy to see you go, you've lost an opportunity to make contract-hunting easier.

7. Don't talk money with the captives: No client wants his employees to know how much he is paying you, because it just starts them thinking: "Why doesn't he pay me more, if he can afford $40 an hour for this guy? Maybe I should quit and work contract." It is likely to lower morale, and won't make your client happy, so be careful. Of course talking money with other contractors is essential, as I've explained elsewhere in this book, but be careful who is listening!

8. For the reasons just explained, don't encourage captives to break free: Your client presumably wants to keep his employees, and isn't going to be happy if your images of the happy-go-lucky freelancer's life persuade them to try it for themselves.

16
The Great Overtime Debate

There is an interesting debate going on in the technical service industry about overtime rates. By law certain employees must be paid one-and-a-half times their regular hourly pay when they work more than forty hours in one week. The law defines two types of employee: exempt and nonexempt. If you are nonexempt, you are not exempt from the law (i.e., your employer must pay you the special overtime rate). If you are exempt, your employer does not have to pay this special rate.

A nonexempt employee is one who is paid hourly, or who receives a salary but does not fit within one of the special exemptions (the Executive, Administrative, Professional, and Outside Sales exemptions).

Many, if not most, agencies do not pay overtime rates. You get paid by the hour, the same amount for the fiftieth hour as you get for the first.

The rest of this chapter comprises two articles I originally wrote for *PD News*, the technical service industry magazine. You can find the address and subscription information for *PD News* in Appendix A.

"We Always Pay Time-and-a-Half for Overtime"

. . . even if the client refuses to pay the extra." How often do you hear that from an agency? Not too often probably, but you may be hearing it more soon. I heard this about a month ago from a representative (I'll call him Mike) of a large national technical service agency. But that wasn't all Mike told me: "It's not a matter of being nice, we *have* to pay it. It's the law."

That is not a belief shared by many agencies, if my experiences and those of my friends are anything to go by. Most agencies do not pay time-and-a-half for overtime; they believe contracting is a different form of employment relationship that doesn't warrant extra pay for overtime, and, more important, they believe they don't have to pay. They may be wrong.

Before I go any further let me state that this isn't agency-bashing. I don't know if contractors *do* deserve higher rates once they have worked forty hours in a week. I've a feeling most probably don't. I simply want to discuss a problem that the industry is going to have to come to terms with eventually.

Quite simply, the law states that most technical service contractors are nonexempt employees, and as such have a legal right to be paid one-and-one-half hours' pay for every hour they work over forty hours in one week.

As an employee of the Labor Department told me, "The assumption is that all employees are nonexempt unless it can be shown otherwise." What does nonexempt mean? It means the employee is covered under the Overtime provisions of the Fair Labor Standards Act. In some situations an employee may be "exempt" from the law (i.e., not eligible for overtime pay), but most people are not; they are "nonexempt."

The first thing to consider is whether a contractor is an employee. If an agency takes out the contractor's taxes and uses a W-2 form to account to the IRS, then yes, he is an employee. (If he isn't paid in this manner he may still be an employee, but that is another article.)

The next thing to consider is the form of payment. If the contractor receives hourly pay, he is nonexempt. No need to go any further, that's it. Hourly-paid employees are legally entitled to overtime rates.

Now, you may have heard of the Executive, Administrative, and Professional exemptions. They are complicated and confusing, but they only come into play for salaried people. Except for outside salespeople, doctors, lawyers, and teachers (specifically including flight instructors for some reason), all the exemptions are dependent on the employee being paid "on a salary or fee basis."

Of course most contractors sign a contract that specifies the rate they will receive. By signing the document aren't they waiving their right to overtime rates? No. The law states that "overtime pay may not be waived by agreement between the employer and employee." In other words, even if the employee doesn't want a special overtime rate, the law says the employer must pay it.

Many agencies now pay "salaries." At least they call them salaries. I once worked for one of the world's largest tech services agencies; I won't name it, but it paid me one of these pseudo-salaries. This is how it works: You agree to a yearly salary and sign a contract with that yearly rate. The agency divides the salary by 2,080 to arrive at an hourly rate. You then report the hours you work and receive that number of hours multiplied by the hourly rate. If you only work thirty hours in a week you are paid for thirty, and if you work eighty you are paid for eighty.

Is this a salary? Probably not. The law says that a salaried employee "must receive (subject to certain exceptions) his full salary any week in which he performs any work without regard to the number of days or hours worked." The employer could deduct money "if an employee absents himself from work for a day or more to handle personal affairs" for example, or, in some situations, when the absence "is due to sickness or accident." But if the employee took the afternoon off, the employer presumably couldn't deduct money (or it wouldn't be a salary). The employer can't deduct money for time spent on jury duty, military duty, or when "there is no work available." (The agency I worked for didn't pay if a contractor was unable to work due to the client company closing for a holiday.)

All this makes the pseudo-salary look more pseudo than salary. There may be some loopholes that a clever lawyer could crawl through, but more and more agencies have decided that there aren't any. Even if the agency can defend its "salary," it has only won half the battle. Just because someone is paid a salary doesn't mean he is exempt. The law specifically mentions computer programmers, for example, and states that *some* "have duties . . . which qualify them for exemption." Only some.

What's the answer? Many agencies and clients no longer allow overtime. (This may not be sufficient, though. The law says that "an announcement . . . that no overtime will be permitted . . . will not impair the employee's right to compensation for the overtime work"—a wonderfully ambiguous statement that Washington should be proud of.) Other agencies pay it, and some are even losing money on overtime. Perhaps agencies will start figuring overtime into the projected costs, paying slightly lower hourly rates to make up for it. But perhaps the law should be changed.

Congress didn't write the Fair Labor Standards Act for us. They wrote it for people on $3.35 an hour, not $33.50. Is a programmer making $80,000 a year being exploited if he doesn't get extra for overtime? I don't think so. The law was intended to protect poor people, not to provide wealthy people with a windfall profit.

The Fair Labor Standards Act is not a new law, but the Department of Labor seems to be applying it to the technical service industry more often now. Many companies are at great risk, potentially owing millions of dollars in back pay; these companies are going to realize how much they have to lose, and start dealing with the problem. "There is no way out," Mike told me. "Our lawyers have looked at the overtime issue, and all they can say is, 'Pay it.'"

After this first article appeared, someone from an agency wrote to *PD News* to say that agencies *can* pay people straight time, because contractors came under the professional or executive exemptions; I decided to go into more detail in a second article.

"Overtime, Once Again"

Robert Marmaduke, of APEX Technical Services in Washington, believes that I incorrectly interpreted the Fair Labor Standards Act in my "Overtime" article of August 21. Mr. Marmaduke stated that employees in "non-trade executive, administrative, and professional positions, spending more than 50 percent of one's day in responsible charge or doing professional design," are exempt from overtime rates (*PD News*, Oct. 2). Perhaps I did misread the Act—it's a little confusing in places, and I may not have seen all the relevant documents—but let me explain why I don't believe I did.

You can find the exemptions Mr. Marmaduke wrote of in the Department of Labor WH Publication 1363, *Executive, Administrative, Professional and Outside Sales Exemptions Under the Fair Labor Standards Act*. This document states that employees' exemptions depend on two factors: "(1) their duties and responsibilities, and (2) (except in the case of doctors, lawyers, teachers, and outside sales people) the salary paid." The document then lists the "tests" for the different categories and states that "*all* the tests must be met" (emphasis in document). The last test in each category is the salary test. For example, the Executive exemption depends on the employee being paid "on a salary basis at a rate of at least $155 a week. . . ." In other words, if an employee is paid hourly, it doesn't matter what his other duties are, he is nonexempt.

The law also provides "special provisos" that effectively bypass the tests. The proviso Mr. Marmaduke describes applies only to Executives, and also requires that the employee must be "regularly directing the work of two or more other employees," a requirement that excludes most contractors. However, there are also special provisos for Administrative and Professional employees, provisos that are less restrictive than the Executive proviso. But if you refer to the text of the regulations (see *Regulations, Part 541: Defining the Terms "Executive," "Administrative," "Professional," and "Outside Salesman,"* WH Publication 1281), you find that in each case the proviso specifies that the employee must be paid on a "salary or fee basis," which leaves us just where we started from.

In his letter Mr. Marmaduke stated that "an agency which offers salaries on a prorated hourly basis in lieu of wages may thus escape overtime [rates]," suggesting that if an employee is exempt the employer may pay an hourly rate. This is the wrong way round, though, as the method of payment is one of the determining factors for exemption status. Thus, an employer cannot say, "I don't have to pay overtime on your hourly rate because you are exempt" because there is, by definition, no such thing as an exempt employee paid an hourly rate. Most contractors receive an hourly rate without the agency calling it a salary, so most contractors do not fit the special provisos that Mr. Marmaduke is relying on.

So the only real ambiguity arises when an agency tells the employee they pay a salary, and then pays what appears to be an hourly rate. Because an exempt employee must be paid a salary, the question hinges on what you call a salary. What I call a pseudo-salary, and Mr. Marmaduke calls a "salary on a prorated hourly basis in lieu of wages," may or may not be a genuine salary. If it is a real salary the contractor, if he meets the other requirements, may be exempt, and not eligible for the higher overtime rate. If it is not a salary the contractor *cannot be exempt*.

In my "Overtime" article I wrote that the pseudo-salary I had received was "probably not" a real salary. I believe it was not a salary. Why? Because it operated like an hourly rate. Being told it was a salary by my employer didn't make it so, any more than being told a duck is a dog changes the animal's species—if it walks like a duck and quacks like a duck, it probably isn't a dog.

However, I'm not a lawyer. There could be occasions when a duck may be legally defined as a dog, in the same way that South Africa occasionally confers the title of "Honorary White" upon blacks. Also, there may be some loopholes that agencies can crawl through.

For example, let's say you receive one of these pseudo-salaries, and one week you work only thirty-two hours because the client closed the office. The company you work for pays you the full forty hours, but subtracts eight hours from your vacation pay. Because no law states that an employer must provide vacations, this pseudo-salary is still a real salary. What happens if you miss work once you have used all your vacation? If your employer pays you less than the normal forty hours, they have probably stepped over the boundary into an hourly rate. (If you receive a salary there are two types of lost time: time your employer must pay for—such as when you are available for work but your employer is unable to provide work for you—and time that the employer does not have to pay for. I'm assuming in these examples that the lost time is of the first kind.)

Another thing to consider is what the agency does if you are between contracts. Agencies that pay pseudo-salaries usually claim that you are their "employee" in all senses of the word, and that they simply hire your services out to various clients. In other words, the relationship should continue from client to client. I would think that if the agency dumps you when you come to the end of a contract the agency has again stepped over the boundary into an hourly rate. However, it may count in the agency's favor if it continues paying you for forty hours a week even when it has no client for you. The agency I worked for did just that. While I worked for them several people were between jobs for long periods of time, a few months in some cases, I believe. I have no idea what bearing this would have on the employees who were never between jobs, employees who worked for a long time with just one client, or went directly from one to the other. I worked for the agency for only six months, and when I ended the first contract, they had an immediate contract lined up for me (I left the day my first contract ended). Does this mean the agency had paid me an hourly rate while I worked for them, or was I paid a salary because *other* people were paid while not working? I imagine each employee's case would be examined separately, but I don't know for sure.

I think that most pseudo-salaries would not stand up to close scrutiny, loopholes not withstanding. What happens if you work a lot of overtime, for example? The agency normally pays you for the extra time, at "straight time" rates, right? But if you receive a salary you would not automatically receive extra pay for overtime—isn't that one of the reasons you gave up "captive" work in the first place?

By the way, any of you who wish to dig further into this quagmire should refer to the following documents, besides the ones I mentioned earlier: *Employment Relationship Under the Fair Labor Standards Act*, WH Publication 1297, and *Overtime Compensation Under the Fair Labor Standards Act*, WH Publication 1325. If you call the Department of Labor, ask for the Wage and Hour Division.

Would I turn down a contract that didn't pay time-and-a-half for overtime? No, at least not for that reason. I would consider the rate and how much overtime I was likely to work, calculate the weekly income, and compare that with other contracts. After all, time-and-a-half for overtime may not be worth much if the base rate is low and you're not going to do any overtime.

Since I wrote these articles the law has changed slightly: Programmers and systems analysts are no longer nonexempt employees if their hourly pay is over 6 times the minimum wage. This change was made with Public Law 101-583 (11/15/91). Section 2 states that the law will apply to "computer systems analysts, computer programmers, software engineers, and other similarly skilled professional workers. . . ." This is rather ambiguous. Are technical writers "similarly skilled professional workers"? Some might say yes (though most programmers I know would scoff at the idea!). However, at the time of writing, the Department of Labor regulations make it quite clear that these regulations *do not* apply to technical writers (see page 8250 in Volume 56, issue 39, of the *Federal Register* [dated 2/27/91]), although the regulations could change.

17
Preparing for Step Two

You've finally found a good contract. You are earning considerably more than you used to earn, and you know what? You don't feel much different from the way you used to. You still come in to work each day, work eight or ten hours, and go home. It all feels much the same as being a permanent employee—until payday, that is.

So where do you go from here? There is no rush, of course. You are already making good money, but probably not as good as you could make. It's funny, but when you first get a contract you're amazed how good the money is. Then, after you get used to it, you start thinking, "Well, maybe I can get $28 an hour." Perhaps you find another agency contract and you do manage to up your rate a dollar or two, so you are making $28 or $29 an hour. That feels great until, a month or two down the road, you start thinking, "I know people making $32 an hour. I ought to start looking for more money." And $32 an hour does feel good, but after a while you start thinking it's not so much, and wondering if you can increase the rate a bit. I know this happened to me, and I've learned from talking with other contractors that most go through the same process.

I suppose it's the very nature of contracting that encourages these thoughts. The contractor is a mercenary; he has no "company loyalty," no hopes for a vice presidentship or a gold watch when he retires. He doesn't believe in company pension plans and company picnics. So what is left? Money. While the permanent employee knows his salary is severely restricted, the contractor is playing a game in which his pay is the prize. Every six months or year he throws the dice again, and sees how much he can make.

One contractor friend of mine has had four contracts in the last year, and each time he moves, his rate goes up. This isn't always possible to do, and it helps if you have experience and a good reputation. One way to make more money, however, is to cut out the middleman, to move to Step Two and find your contracts yourself, without the aid of an agency.

Unfortunately, finding work without the agencies is more difficult. The agencies have full-time salespeople who know hundreds of companies in your city. Those companies often call the agency looking for contractors. Further, many companies will not do business with independents. Some have large staffs of contractors and would rather let an agency handle all the paperwork and just present them with one bill every couple of weeks. Others want to avoid the legal problems I explain in chapter 21. So you have to put more effort into finding independent contracts. On the other hand, you will sometimes run into companies that prefer not to do business with agencies, perhaps because they have had problems with them in the past.

How do you begin? The chapters in the next section will describe what you need to do. It's quite simple, just a matter of building your reputation and network and knowing how to look for contracts. The salespeople in the agencies do it, so you can, too. It may take six months or a few years to prepare for Step Two (it took me one year from getting my first agency contract to finding an independent contract), but you can begin as soon as you start your first contract. There are two main things you must do to prepare yourself.

Build Your Business Capital

The first thing is to begin saving money. I know it's nice to have all that extra cash; if suddenly you go from earning $30,000 to making $50,000 it's tempting to start living a $50,000 lifestyle. But the extra $20,000 is not profit—you are in business, remember, and a portion of that extra money should be earmarked for "business expansion," to allow you to move on to the next phase of your business. Of course you can spend a bit more than you used to: take a longer vacation, go out to a good restaurant a bit more often. But don't make the mistake that most contractors seem to make.

Contractors get carried away with all the money. But if you don't save enough money to live on for a few months, you are going to hurt yourself in a couple of ways. You are going to make it difficult to move on to Step Two, in which you can make considerably more money, and you can end up in a position in which you have to take a low-paying contract because you don't have the time (i.e., the money) to find a better one.

Why do I say you need a few months' living expenses saved? It is not just because you may be out of work for a few months—you may, I suppose, although many contractors are rarely out of work for more than a week or two—but for psychological reasons. If it takes two weeks to find a contract, it's no good having four weeks' living expenses (two weeks while you look for a contract and two weeks waiting for the first paycheck). After all, you don't know how long it will take to get a contract, until you finally sign one. What happens if you are offered a mediocre contract after one week of looking? Do you wait, or do you think, "I'm going to be in real financial trouble if I don't get work within a week," and take the offered contract? I have seen contractors in this position before. I have one friend working for $22 an hour, when he could easily get $28 an hour—all because he didn't save.

If you do save, you have more confidence. You don't think about financial trouble if you know you can last three or six months if you need to. Not that you want or expect to wait that long, but knowing you could reduces the psychological pressure.

You need money to move on to Step Two. In Step Two you are running a business. You will have to invoice the client, and then wait for the client's accounting department to pay. On my first independent contract it took six weeks from my first day at work to depositing my first check in the bank. But it's not uncommon for companies to take sixty or even ninety days to pay invoices, even if the terms are actually net thirty. (Although strictly speaking you are simply a supplier, some clients will be sympathetic and try to get the money to you as soon as they can.)

When I got that contract, I went straight from an agency contract to the independent contract with only the weekend off between. Imagine if it had taken two or three weeks to find a contract—eight or nine weeks without an income. That is why you must save money. If you don't save, you will be unable to take independent contracts, or be in financial trouble if you do.

Educate Yourself

The other major preparation for Step Two is to educate yourself. Talk to people about the contract market, about who hires contractors and what they pay. Talk about what the other contractors make and what the range of contract rates is in your profession. Working through agencies can help you educate yourself because more often than not you are thrown in with other contractors: Draw on their experience and learn from their knowledge of the local market. Join their little network groups, often nothing more than a group of contractors who get together for lunch every few weeks. Ask about local professional societies and associations.

Most important, keep records. Keep a card file with the names and telephone numbers of the people you meet: contractors, non-contractor colleagues, employers. When you go on job interviews through agencies, keep business cards. You will build up a file of names, of people you know and people you don't, a list of names that can help you find a contract when you need one.

IV.
The Second Step in Freelancing

18
Networking

Eighty percent of success is turning up.
—WOODY ALLEN

Networking. The very word frightens many people. They think of endless hours at seminars and association "events," forcing themselves to smile at people they don't know or even want to know. They think of pushing themselves upon strangers, asking for business cards and going into long explanations of what they do and why the stranger should consider hiring them, or tell a friend to hire them, or tell a friend to tell a friend.... Many people think of networking as "using" people, and in a way I suppose it is. But it doesn't have to be calculating or dishonest. My friends use me when they are looking for work, and I use them when I am looking. I don't socialize with people I don't like, so the people I go for lunch with or have a beer with after work are not being used—I would be there even if they were of no "use" to me.

That is one of the great misunderstandings about networking, that you have to socialize with people you don't like so that you can get something from them. I regard networking differently: As you naturally form friendships with people you work with, why not use each other's knowledge to help each other?

Networking doesn't have to be hard. To me networking is having lunch with a group of friends—friends who also happen to know what is going on in the job market. There are certain networking techniques that you may not want to try, but that doesn't mean you can't build a network. If you are a normal, sociable human being, you can build a network just by making friends.

First, though, what is it you want from your network?

The Information You Need

There are several things you need a network for, and looking for an immediate contract is only one of them. You want to get a "feel" for your market. You want to know names of companies, other freelancers, and managers. You want to know which companies are hiring, and which are laying off. Who pays well, and who doesn't. Keep a card file, or a database on your computer, and add this information periodically.

You want to know how much people make. This is sometimes a sensitive question, especially with people you don't know very well. Make the effort to get this information, nevertheless. Ask friends—no one I've asked has seemed to take offense, and you'll often find friends volunteering the information. If you are not quite sure whether you know someone well enough to ask him, ask him anyway. Remember, *Who fears to ask, doth teach to be denied.* In other words, if you are too modest to find this information, you will be denied the best rates.

Finding out what people are making tells you a few things. It tells you what your potential is—if you know less qualified people than you making 20 percent more than you, figure you can make at least 20 or 30 percent more than you are making now. You can get an overview of the market. What is a good rate or an average rate? You can't tell if you know only how much you make, or what one or two other people make.

You also will find out which companies pay the best rates, and will know how much to ask a company for if they offer you a contract. For example, I know companies that simply don't pay more than $25 per hour for technical writers, but I also know of companies that will pay $37 or $38.

You also want to know what companies are like: which ones are fun to work for, which ones take a long time to pay (or never pay), which ones are likely to be hiring soon, and which ones are in financial difficulty.

The information you need is anything that will help you find work and get a good rate. You may not need work now but you still need your network, because you can't build a network in a short time, and you can't learn all you need to know about the market in just a few weeks.

Building a Network

So here are a few tips for building a network. Use the ones you like, discard the ones you don't.

1. Keep in contact with "key" people: You will find contractors who seem to know what is going on just about everywhere. They have networks that keep them informed of new contracts, of companies that may need people soon, of contractors who are about to quit (even though the client doesn't yet know it), of companies that may hire people but are just as likely to cancel the contract early, and so on. Contacting one of these key people is like building an instant network.

2. Stay friendly with some of the agencies: Even though I try to avoid working through agencies, I still have a few friends in agencies who feed me information—not necessarily information that will get me a contract immediately (although one agent did give me a few hot leads, and he didn't have anyone to fill the contract, so it was no loss to him), but information about companies and other agencies that may be good background data.

3. Keep a card file and put a card in file for everyone you work with, either in your profession or in related jobs, anyone who may have information that could be useful to you: Note the name, telephone number, company, and position, and any other relevant information, or staple a business card to the file card.

4. Keep company telephone directories when you leave a contract: You may need them later to contact someone.

5. Put a card in your card file for every company that interviews you: Staple a business card to the file card. Note how the interview went. Keep this information even when the interview was set up by an agency. The information could still come in useful later. A couple of companies that originally interviewed me through agencies later offered me work. This is advice that I didn't use early enough—I didn't keep records of interviews, so I lost a lot of good information.

6. Join professional associations: For example, the Dallas chapter of the Society for Technical Communication has over 300 members, all people in companies that use technical writers, an excellent source of contacts. Dr. Jeffrey Lant, in his book *The Consultant's Kit*, suggests that you go to professional association meetings and "go to the bar to meet people. Invest some money in buying drinks for people." You don't need to be so calculating about it; just go to meetings and make friends.

7. Even if you are not a member of a local association, see if you can get hold of the membership list: These lists can help you when you are contract-hunting.

8. Find out which companies are associate members of local associations: These names tell you who hires people in your industry and profession.

9. Go on job interviews even if you don't need work: Jeffrey Lant also suggests going on interviews for captive positions and not revealing that you are a contractor, so that you can find out about the company and make contacts. This may be a good idea in some circumstances, but I wouldn't do it indiscriminately. I once went on an interview after an agency had submitted my résumé without asking me; I already had a contract, though, and didn't want to leave. It was time well spent, though, because we agreed to keep in touch and discuss a contract later.

10. Find the local "job banks": The Society for Technical Communication has job banks in most large cities, and other associations do also. The leads may be for captive employment, for independent contracts, or for agency contracts. Try to get all the leads; some job banks will try to screen people, so that you get only the leads that they think are suitable for you. There are a couple of problems with that: Even if a company tells the job bank its list of criteria, it may settle for less if it can't find the right person, or it may settle for slightly less experience in the specified area if the person has more experience in a related area. And just because you don't fit the criteria doesn't mean that the company hasn't got other positions more suited to you, or will not have soon.

11. Check your local paper's classified ads each week for companies seeking people with your skills or associated skills: The companies may be looking only for captives, but it doesn't matter—they may not be able to find a captive, and at least you have the name of a company that *may* need freelancers one day. For example, I collect ads from companies looking for technical writers, so that when I'm looking for work I know who to call. The ads sometimes even have names of the people who handle the hiring, which can save you a lot of time when you are trying to get through to a company.

You may want to send a résumé and business card, with a cover letter explaining that you are a freelancer, to all the companies advertising for your profession. Some of these résumés will be kept on file for later use. It's a simple way to build "name recognition." If one of these companies is looking for a contractor at some point in the future and again receives your résumé, it may recognize your name, and is more likely to read your résumé.

12. Go to job fairs: You can hand out your résumé at job fairs. Include a cover letter explaining that you are a freelancer, and a business card. Most companies keep résumés on file, and although companies at job fairs are usually looking for captives, who knows what their needs will be in the future?

13. When someone in your network goes for a job interview, ask them where, and with which company, and whom they spoke to: Ask what the interviewing company was looking for, and how much they were offering. Put this information in your card file.

14. Get to know the personnel recruiters at the companies for which you work: Keep their names and numbers. One personnel recruiter even gave me a list of recruiters at dozens of local high-tech companies throughout the area.

15. Business lunches are a great way to meet people in your profession: I occasionally go for lunch with friends I met while working on previous contracts. As we all move from contract to contract we meet different people, expanding our network as we go. We invite new friends to lunch, and introduce them to the other members of the network. This is beneficial to all of us, both new and old members of the network.

16. Have parties: As a freelancer you can introduce friends to each other, people in the same profession working in different companies. Why not invite friends you have met at different contracts to a party? Not only is it fun to have parties with new people (instead of the same old crowd), but you are doing your friends a favor.

17. Keep in touch with people: If you move, send change of address notes to companies and friends. I made the mistake of not doing this when I moved, and lost a contract; a company that had offered me a contract six months previously (and then backed out because a senior VP had said no) was now ready to hire—but didn't know where I was! Even if you don't move, you might want to do what many salespeople do: send notes or cards to potential clients now and again, to make sure they don't forget you.

18. Get to know other people in the company you are working for: Other departments also may hire people in your line of work, so get to know other department managers.

A note about business cards, by the way. I always forget mine, and rarely give them, which is a mistake. However, so far as business cards go, *it is more important to receive than to give!* Most cards you give away will be lost or trashed within a day or two (that's no reason not to give them, though, because now and again they *will* be saved—and maybe used one day). On the other hand, the cards *you* receive should be saved. Those cards are an important part of your network, and shouldn't be wasted.

19. Help people: It's important to help people. If this comes naturally to you that's fine; if not, make an effort. When friends are looking for work I do all I can to help them. I give them the names of people looking for freelancers, information about companies that may be hiring soon, and my list of technical service agencies if they want to mail résumés to the agencies. I don't do this because I expect anything in return—I would do the same if I was about to leave the area and never see these people again. However, I know that because I have helped my friends, they are apt to go out of their way to help me when I'm out looking for a contract.

You should even help the technical service agencies. If an agency calls you when you don't need work, don't just hang up. Ask about the job—how much they are paying, where the job is, what type of job, even the company name (although they may not tell you). This is all good information of course, but you also should try to find someone to fill the slot. Don't spend too much time on this, but if you know someone who is looking or may want to move to a new contract, get them to call the agency, and ask them to mention your name.

One of my technical writing friends always helps the agencies. He pulls out his little book of addresses and goes through it looking for the names of people who may be interested. As a result, he gets a lot of calls from agencies looking for contractors, and agencies throughout Dallas know him, something that can come in handy when he is looking for work. The person getting the job is grateful also, and will undoubtedly help him the next time he is looking for a contract.

A final note. Like the CIA, you are going to collect a lot of useless information. Useless in the sense that most of it will never lead to a job. Only a small fraction of your information will help you find contracts, but you don't know which pieces—if you did, you wouldn't need a network. You don't know where the important information is going to come from, so the more of these methods you use the better.

Salespeople understand that you have to make a lot of contacts to make a sale. Most of those contacts are useless. Most will not lead to a sale. But the salesperson doesn't know which contact will buy, so he has to contact ten, or twenty, or fifty people for each sale. You should remember that you are a salesperson, selling your own skills and services.

You can think of networking as boring and tedious, but I think of it as a kind of game. It's fun to gossip about what's going on in the market, and it's amusing to see people's faces when you give them the inside scoop from a company they have applied to. It helps if you enjoy gossiping, of course—most people do—because that is what networking is all about.

19
Looking for Work

So, you're ready to look for an independent contract. You have money in the bank, and a good network. How are you going to start? The chapter describes several ways to search for work. I suggest you start at the top and work your way down the list. With luck and a good economy you may not get too far before you find a contract. If you are working on an agency contract now, try beginning your job search about six weeks before the contract finishes, which should give you enough time to find a new one. Of course, if you are actively networking, you may run across a contract offer even before you start looking for work in earnest.

You are searching for leads, information that will lead you to a job. When you find a lead, act on it immediately, before moving to the next step in the list or the next person in your card file. There is nothing more frustrating than hearing a prospective client tell you that you have just missed an opportunity.

When you talk to a prospective client, don't just ask if the client is looking for someone with your skills, but ask if the client knows of anyone else who may be hiring, or if someone else in the company also hires people in your profession. Note down as many names and numbers as possible. *Don't forget to ask these questions*. You can use this method to multiply your leads, one lead bringing one or two other leads, just as a professional salesperson does.

After speaking with a prospective client, send a résumé and a thank-you letter, and a business card if you have one. Even if you don't have a chance of being hired right now, the résumé will help the client remember you in the future, and the client may even keep the résumé on file.

The Steps to Your Next Contract

1. Call the "key" people in your network, the people who seem to know what is going on everywhere: These people can often give you several good leads each.

2. Call the rest of the people in your network: friends, colleagues, people you worked with some time ago, and so on.

3. Call companies you have worked for in the past, and people whom you have "almost" worked for (people who have interviewed you or offered you work in the past): These people already know you, so if they are looking for someone, they can save themselves a lot of time and trouble by hiring you. If the person who hired you has left the company, try to find out where he has gone; he may have transferred to another managerial position at another company.

4. If you are working on a contract, ask your client if any other departments in the company hire contractors: You may be able to transfer to another department, but remember that if you are working through an agency, the contract you signed—and the one signed by the client—may stop you from working for the client without the permission of the agency. This clause may not be legal in some states, but still, it may be enough to scare off the client.

5. Call the professional societies' job banks and ask who is hiring: Don't just ask for contract work—often companies looking for permanent employees may not be able to find anyone suitable to take the job, or may be hiring contractors as well.

6. Check the newspaper classified ads for companies looking for your skills: Again, don't just check with the ones advertising for contractors, but even those looking for "captives." Call these companies even if they say don't call—you can tell them that a friend told you about the job and that you haven't seen the classified ad. Even if the company address is a post office box, you may be able to track them down. Your library should have a directory—called a "reverse" telephone directory—that allows you to look up an address to find a telephone number; these directories also list post office boxes and whom they belong to. Of course, if the company recently rented the box, or rents a box from the newspaper, you won't be able to find out who it is.

7. Go through the classified ads you have collected in the past and call the companies that were hiring people in your profession: They may have something available soon.

8. If you have a computer and a modem, try advertising your services on a bulletin board: For example, CompuServe has bulletin boards that you can advertise on (all you pay is the connect fee), and with 500,000 subscribers enough people will see your ad to make it worth a try. CompuServe is a national service, but you may be able to find a local bulletin board.

9. Get hold of a list of other people in your profession and call them: I have a list of about 300 technical writers in the Dallas area, given to me free when I joined the STC. While I have not yet used this list of names, if I had got this far without finding a contract I would definitely start calling these people.

Admittedly, this is not easy. Most people find "cold-calling" distasteful, if not downright uncomfortable, but it isn't so bad. It's not as if you are calling people to sell aluminum siding or encyclopedias—the type of telephone work that leads to incredible rejection rates. No, you're not *selling* anything, and you probably don't even have to call at home—80 percent of the people on my list gave their work numbers. Though I haven't used the list, I don't believe it would be difficult, and I don't believe I would get much rejection, if any. (I have done telemarketing before, at a low point in my career, so I know what it's like.)

I would simply call and introduce myself, explain that I'm a fellow technical writer, that I found the person's name in the association's list, and ask if he had a moment to speak. Most people will say yes—after all, what would you do? There is no threat, this is a fellow association member, it's only natural that you would take a moment to talk. I would then explain that I am a contractor looking for work, and ask if he knew of any companies looking for contractors right now. You may get a lead right away, but even so, keep asking questions. Does he know of any company looking for permanent employees? Does he think his company may require someone soon, contract or permanent? Does he have any friends in other companies whom you could call and ask? (Remember to add these names and numbers to your list.)

I believe that the response you receive will be very favorable. People enjoy talking to others in their profession, and people especially enjoy finding a reason not to work for a few minutes. People usually enjoy helping others, sometimes because they hope someone will do the same for them someday. Of course some of the people you talk to will be unfriendly, and you won't get very far with them. You will know from their tone of voice when not to bother asking more questions. Just thank them for their time and go on to the next person in the list (make a note by that person's name not to bother with him next time).

If you talk to enough colleagues you will get a clear view of what is going on in your market and find dozens of leads. You also will make many useful contacts for the future. And if someone is especially helpful to you, send him a thank-you letter, and include a résumé and a business card.

10. Find a list of companies that can use your services: This is very easy to do. Go to your library and ask the librarian for help. You should be able to find a directory in a book, on microfiche, or on computer disk, that will list all the companies in your area. Your library may have Microcosm, which is a microfiche directory, or The Business Index, which is on a rotary microfiche machine. One advantage to Microcosm is that you can use a microfiche copier to copy pages; with the Business Index you need to write the names and addresses down.

You use these directories to pick the companies by zip code, name, size, and so on, in order to limit the search to the most likely candidates. Note the company name, phone number, and officers' names if included (many of the lists include the president's, vice-president's, and treasurer's names, which may be useful for smaller companies). You also could use the Yellow Pages to find company names, just calling every company in a likely category, one by one. You will get a lot of companies that are not likely to be good prospects, but on the other hand you will save a lot of time in research.

Now call these companies. This is more difficult than the telephone work you did in the previous step, and you will get more rejection. Call the switchboard and ask for the head of the department that is most likely to hire you. For example, ask for the documentation or publications department, or, failing that, the engineering department. Or call the human resources or personnel department, and ask them which departments hire technical writers.

Some companies hire contractors through the human resources department, while others leave it up to department managers to find their own contractors. Try to find out how the company operates, so you know whom you need to talk to. You may want to read *Freelancing—The First 30 Days*, by Bill Coan, for a detailed explanation of cold-calling (see Appendix E).

11. Go to the local job fairs, talk to employers, and spread a few résumés around, with cover letters and business cards, explaining that you are a contractor.

"We Only Use Agencies"

Some companies will tell you that they work only with agencies, that they will not hire independent contractors. They do this for a variety of reasons, not the least of which is that they don't want trouble with the IRS. Don't give up yet. Tell the client that you would be prepared to work through an agency. Explain that if he interviews you and decides to hire you, you will find an agency acceptable to him, and work through that agency. Explain that you have contacts in several agencies, and that you can easily find a suitable one. (Many companies have a list of agencies that they work with.)

This ploy may not work, but if you have a reputation in your area, or skills or experience that the client is having trouble finding, you may be able to get them to agree. And if the client does decide to hire you? Negotiate a rate high enough to pay you a good rate and pay an agency's overhead and profit. Go to several agencies and "sell" the contract to them. You should get a high rate, because you are bringing the work to them. All the agency has to do is process the paperwork to make the money. You have already done the most expensive, time-consuming part of the agency's work for them—finding the contract and lining up a contractor.

Of course you need to get your sums right. The rate the client pays must be enough to pay you a good rate, with enough left over for the agency to pay your taxes (10 to 15 percent) and give them a small profit. Clients are likely to pay a higher rate to an agency than they might pay an independent, anyway, because they understand that there are more people taking a bite of the pie. For example, even though $38 per hour is a good technical writing rate for an independent in the Dallas area, companies often pay agencies

$42 or $43 an hour. Remember that your cut doesn't need to be as high as a true independent contract rate; you will pay less FICA (7.65 percent of the first $53,400 of income, instead of 15.3 percent), and the agency may have benefits you can use, such as medical insurance. On the other hand, you will not be able to deduct most of your business expenses, and you won't be able to set up a self-employed person's pension plan. You may be able to find an agency that will pay you on a Form 1099 so you can retain your independent contractor status (but make sure you read chapter 21 first).

Here are a few more things to consider. First, don't count your chickens before they hatch. Don't stop looking for work just because it looks as if you are about to sign a contract, because the contract will often fall through (remember Sam Goldwyn's comment about verbal contracts not being worth "the paper they are printed on"). Whatever you do, *don't* tell someone you are unavailable for work until you are absolutely sure, until you have actually begun working on the contract. If you tell them you are available and then the other contract comes through, you will have to tell the second contact that you are unavailable, but that's business. Few companies would be upset, because they know that until a contract is signed anything can happen; they could also pull out at the last moment (and often will).

Also, don't be discouraged when people don't call you back. It is hard to take rejection, and it's easy to see rejection when it really doesn't exist. You may think you have a contract coming up any day and never hear from the company again. That's not necessarily an indication that the company has decided that you are not good enough for the job. Perhaps your contact at the company has quit or been laid off; perhaps the company has financial problems, or lost a contract that *your* contract depended on; maybe the department didn't get approval to hire you, or has been told to wait six months. There are thousands of reasons for people not calling back. It would be more polite for them to do so, perhaps, but it's not necessarily a reflection on your skills or reputation. Don't worry about it, just keep calling; if you use all these methods and there is work to be found, you will find it.

20
Taxes for the Freelancer

Note: The tax information in this chapter is related to the 1991 tax year, unless otherwise noted. Although these figures and the laws discussed may change slightly, this chapter will still give you an idea of how to file your taxes. For more up-to-date information, refer to a current tax guide or to IRS publications.

Most freelancers I know are frightened by taxes. They know next to nothing about how to file their taxes or what deductions they can take, and much of what they think they know is wrong.

Taxes are easy to handle, though, if you are willing to invest a few evenings reading a tax guide. I don't intend this information to be a substitute for a good book on taxes (I use *J. K. Lasser's Your Income Tax*), but it is an overview, a way to give you an idea of what is involved and where to go next. I'm not going to tell you exactly how to fill in every form you may need (I don't have room for that), but I will discuss the areas of the tax law that affect freelancers. I have also assumed that you don't have any employees, so I have not covered tax law relating to employee withholding, employee pension plans, and so on.

Why bother to do your taxes yourself? Why not hire someone to do them for you? There are several reasons. First, if you hire someone to do them you have to give that person all your records, in some semblance of order. Record keeping is 90 percent of the work involved in doing your taxes, however, so you may as well finish off the last 10 percent and save some money.

This is especially true if you have your own computer. You can buy very good, inexpensive programs that compute taxes for you, so why not just plug the numbers into the computer and let it do the job?

Another reason not to use tax preparers is that they often don't know what they are doing. You may have seen newspaper stories by journalists who took the same tax records to ten different tax preparers and got ten different results. Some of my freelancing friends have received incorrect information from their accountants. If you want to make sure your taxes are done properly, you will have to do them yourself.

Another reason to do your own taxes is so you can make sensible financial decisions. If you don't know the effect of a decision on your taxes, how can you decide properly? For example, I know people who have bought houses in order to "save money on their taxes," not realizing that the amount they were saving was more than wiped out by extra expenses involved in owning the house. (I live in Texas at present, where nobody expects their house to appreciate in value for a while.) You may need to decide whether to invoice a client or wait a few weeks to push the income into next year. If the extra income would push you into a higher tax bracket, you might consider waiting a few weeks to invoice the client, and paying a few bills early.

I'm going to begin by discussing certain controversial aspects of the tax law. You may talk about my ideas with other freelancers and find them in total disagreement, or you yourself may not believe me; many people take the attitude "We've always done it this way, so it must be right." I have written about some of these subjects in my column in *PD News*, and have found many readers in agreement, and many saying, "Hogwash." I suggest you read the tax information booklets I mention, and discover for yourself who is correct.

Is Mileage Deductible?

There is a belief among freelancers that they can deduct a certain sum from their taxes for each mile that they drive while on business. This belief is quite correct, of course—in 1991 you could deduct 27.5 cents for each mile driven, or you can deduct a portion of the actual vehicle costs (gas, insurance, maintenance, etc.). The real question is, what is business-related mileage? Most freelancers seem to think that driving to work is business-related mileage. After all, they are freelancers, they say, so they are in business for themselves (even many freelancers who are, strictly speaking, technical service agency employees stick by this reasoning). Some agencies even pay mileage as an extra benefit. For example, if the contractor drives thirty-five miles from his home to the client's office and back each day, the agency would pay the contractor $9.10 (35 miles times 26 cents). This agency regards this amount as a business expense, so taxes are not normally withheld.

Some contractors I know also maintain an office at home, so they reason that driving to work is not commuting, it is traveling from one business location to another. If a contractor worked mainly at a client's office, however, it is unlikely that the IRS would accept such an argument. As *J. K. Lasser's Your Income Tax* puts it, if a home office is not "your principal place of business or a place in which patients, clients, or customers meet or deal with you in the normal course of your profession or business," then it is not a valid office.

Driving to work and back is a "commuting expense," the IRS says, and so it is not deductible. If an agency pays an extra sum based on the commuting mileage, that sum is ordinary income and is itself taxable. What then is the difference between commuting and business-related driving?

The answer is very simple. We all know what commuting is: travelling to work each day. It doesn't matter if you are a captive employee or a freelancer with several different contracts, commuting is commuting. If you work in the same office for several months and travel to and from that same office each day, the IRS regards that as commuting. The IRS states: "You cannot deduct the costs of driving a car between your home and your main or regular place of work, (even if) you are employed at different locations on different days within the same city or general area." Now, if you were to drive to an agency office, pick up an assignment for the day, and then drive to that other location, the drive from the agency to the location would be a valid business expense. This is rare in the technical service industry—more commonly you work at the same location for months or even years.

I've been told that some agencies falsify records, to make it look as if the contractors come into the agency office each day, before going out on assignment—this then allows the agency to deduct the mileage rate paid to the contractor as a business expense, rather than as salary (if it is salary, the agency should withhold taxes, and pay its portion of the taxes as well). Some independent contractors claim that because they have a home-office, any driving they do is deductible. One contractor told me that if the IRS audits him he will say that he goes into his office each morning before driving to work. I don't believe the IRS would accept that; all they have to do is examine his invoices or interview his clients to find that he works in the client's office almost every day, for at least eight hours—in such a case they may disallow the home-office because it is not his "regular place of business."

The IRS allows deductions for only these types of commuting:

1. **If you are out of town on a business trip**—I'm going to discuss what constitutes a business trip in a moment.
2. **If you use your car to carry tools.** However, the IRS lets you deduct only the expense directly related to carrying the tools. For example, if you didn't need your tools, you could pay $1 to go by public transportation. Because you need tools, you have to drive your pickup and incur $3 mileage expense a day. The extra $2 would be deductible.

3. **If you commute to a temporary job location,** even if (since 1991) it's close to your usual place of work. Many contractors feel that this always applies to them, though *why* I can't imagine. If you work for a company on contract for six months, at the client office, then the office is your principal place of work. If the client wanted you to work in another city for a few days, that would be a deductible commuting expense.

The situation is a bit different if you work at home or in your own office and occasionally visit the client, of course. If your home-office is your principal place of business, then a trip to the client once a week would be a justifiable business expense.

Per Diems and Business Travel Expenses

Many agencies pay a per diem, usually to *road-shoppers*, contractors who travel from other areas of the country to take a contract. Per diem means *per day*, of course, but as I've mentioned elsewhere, the technical service industry doesn't seem to know that. If an agency quotes you a rate and then tells you they also will pay a $150 per diem, don't get excited—they mean $150 per week, not per day.

Anyway, most contractors assume the per diem is a nontaxable reimbursement for travel expenses, but most contractors are wrong. While there are some situations in which this may be true, usually the per diem is taxable income.

In response to an article I had written in *PD News* one contractor wrote to state that "when one is working on the road, away from one's home overnight, expenses may be deductible." The writer didn't explain exactly what he meant by "away from home overnight," but if he meant that when someone travels to another town on business and stayed overnight (or even a day or two), then the expenses would be deductible. But if he was referring to the *road-shopper*, the contractor who usually jumps from location to location and has no regular workplace, then the commuting would not be deductible. The following comes from my response, published in *PD News*.

> Most of this talk about how to deduct expenses and report PD (per diem) is irrelevant. Most contractors working away from what they regard as their home are probably not eligible to deduct travel and living expenses. Even if they receive a PD it should be reported as taxable income. Why? Because usually the IRS will regard where you work as your tax-home. I haven't done a survey, of course, but I don't believe most contractors' work situations fit the criteria laid down by the IRS for claiming living expenses.
>
> The IRS intended travel and lodging expenses to be deductible for people temporarily out of town. Most "temporary" stays are a few days or weeks, but could be as long as two years (if, for example, a company you work for asks you to work temporarily in a subsidiary in another town). However, most contractors are not in this position. As *J. K. Lasser's* says, if you "do not have regular employment where you live . . . the IRS will disallow the deduction on the grounds that the expenses are not incurred while away from home: the temporary home is the tax-home."

Many contractors, especially single people, do not maintain houses or apartments in their home towns: "Bachelors will find it difficult to get the deduction because they often do not keep regular residences," says *J. K.* Some contractors have trailer homes, and move their homes with them. In such a case "you move from project to project and you have no other established home . . . each location becomes your principal place of business and, therefore, you are not 'away from home.'" This would apply even if you didn't physically move your home with you, but still traveled from job to job around the country.

Even if you maintain a house or apartment in your home town, you may still not be eligible to deduct expenses. Even if your spouse or children live in your home town, you may still be out of luck, especially if you have not recently worked there.

The IRS considers several circumstances when determining tax-homes. The more of the following factors that fit your case, the less chance you have of getting the IRS to accept your deduction.

1. You have no regular employment in your home town.
2. You do not maintain a home in your home town.
3. You move from job to job in different locations.
4. You have no spouse or children living in your home town.
5. Your present contract is indefinite, or scheduled to last two years or more.
6. You did not work in your home town before the present contract.
7. You don't intend to return to your home town after your present contract.
8. You are not continuing to seek employment in your home town.

Scan through a copy of *PD News* and look at the Length of Contract listings. If a listing is "indefinite" or over a year, the per diem is almost certainly not tax-free (perhaps the agency doesn't withhold taxes from it, but the IRS regards it as taxable). If it says two years or more, it is "considered an indefinite stay, regardless of the facts or circumstances," says the IRS—so per diem would be taxable.

My most recent copy of *J. K. Lasser's* tells me that the IRS will disallow a travel expense deduction if you have no regular job in your home town, but that the tax court has allowed such a deduction. However, an appeals court has overturned the decision, agreeing with the IRS. (Unlike most crimes, in which the state considers you innocent until proven guilty, where tax "crimes" are concerned you are effectively considered guilty until proven innocent. The IRS says you owe them money. If they don't change the assessment during their appeals procedure, you can pay them or take them to court. If you go to court *you* are suing *them*, so the court regards the IRS as innocent until you can prove they are wrong! Thus, an appeals court can overturn a tax court decision and decide in the IRS's favor.)

Does all this mean that you cannot deduct mileage driven to work, or that you must pay taxes on per diems? No. After all, many contractors throughout the country do deduct mileage, and don't pay taxes on per diems, and get away with it for years. What it means is that if you get audited by an auditor who knows the law, you will lose the deduction and may have to pay penalties.

Many contractors get away with bending the rules slightly, or even breaking them completely. They take advantage of every grey area, and the fact that the IRS can't know their exact work situation unless it audits them. So you will find contractors who tell you the points I have raised are wrong (because they've seen so many others deducting these expenses that I must be wrong), or that it doesn't matter, because you won't get caught.

But taking every liberty with the tax system possible may be, as one contractor put it, "killing the goose that laid the golden egg." Too many contractors have avoided paying taxes, or taken illegal deductions, so the IRS—and even state regulators—is out to clean up the mess. The easiest way is to remove the people causing the problem: stop independent contracting, and force the contractors to work as employees of agencies or of the clients. Many individual states have already clamped down on independents, making it difficult to find a company that will risk an independent contract. The agencies have reaped a bonanza, and often support the IRS's efforts. If independents went by the letter of the law there would be no incentive for the IRS to go after them, though this may be wishful thinking—it's probably too late. (But read chapter 21—there are still things you can do to avoid legal problems.)

The Home-Office—How Much Can You Deduct?

The IRS is tightening up on home-office deductions. You can only deduct business expenses related to part of your home if that part is regularly and exclusively used as 1) your principal place of business, or 2) as a place for meeting clients or customers in the normal course of business, or 3) in connection with your business if it is a separate structure unattached to your home. The IRS wants to make sure that you really are using that part of your home for an office, and not just calling the den your workplace so you can save a few more dollars on your tax bill. If you file Schedule C, you must now state if you are claiming expenses related to a home-office: checking the Yes box may be a good way to get audited, especially if you don't have much self-employment income.

Most contractors would not fit the IRS's criteria for home-office use. I even know some contractors who do maintain home-offices but don't claim deductions because it is a "grey area" and they don't want the hassle of an audit. But if you are using part of your home for an office, and fit the IRS's criteria, you can deduct a portion of real estate taxes, rent, mortgage expense, maintenance, utilities, and insurance. See IRS publications 587 and 936 for more information. And, starting in 1991, you must fill in Form 8829 describing your use.

Filing Your Taxes

How you file your taxes depends on your freelancing status. Most freelancers working through agencies are technically employees; the agency withholds taxes and gives the freelancer a W-2 form at the end of the year stating how much it paid him, and how much tax was withheld. If you are an independent contractor—that is, if a client or agency pays you without withholding taxes—you should receive a Form 1099 from each client showing how much money they paid you.

The Freelance "Employee"

As an "employee" of an agency, you file taxes much the same as a captive employee of a company. You must complete Form W-4 for your agency, showing how much tax they should withhold and then, by the middle of April following the tax year, you must file your tax forms. You will normally file Form 1040, 1040A, or Form 1040EZ, plus Schedule A if you have sufficient itemized deductions. The only major difference between the freelance employee and a true, captive employee is that the freelancer is more likely to have business expenses that can be deducted. Remember, though, that the IRS regards both types of employee as the same. There are no deductions that apply to you as a freelancer working for an agency that do not also apply to captive employees.

The expenses you may be able to deduct are "unreimbursed business expenses" and the "expenses of producing income." By *unreimbursed* the IRS means valid expenses that your employer does not pay you for.

Read the following list to get some ideas on what you may be able to deduct. These items are from *J. K. Lasser's Your Income Tax* and IRS publications (see a tax guide for details of when and how these items may be deductible):

Airfares
Auto club membership
Books used on the job
Business machines
Business use of part of your
   home—but only if you use

that part exclusively and on a
regular basis in your work
and for the convenience of
your employer
Car insurance premiums
Christmas gifts

Cleaning costs
Commerce association dues
Commuting costs when out of town, carrying tools, or at a temporary job site
Convention trips
Correspondence courses
Depreciation
Dues to professional organizations and chambers of commerce
Educational expenses if required by your employer or by law or regulation to keep your present salary or job, or for maintaining or improving skills you must have in your present occupation
Equipment
Fees to employment agencies and other costs to look for a new job in your present occupation
Foreign travel costs
Furniture
Garage rent
Gasoline
Gifts
Home-office expenses
Hotel costs
Instruments
Jury duty pay handed over to employer
Labor union dues
Laundry
Legal expenses
Magazines
Malpractice liability premiums
Meal and entertainment expenses
Medical examinations
Parking fees, tolls, and local transportation
Passport fees
Pay from previous year repaid to employer
Physical examinations your employer said you must have
Safety equipment, small tools, and supplies needed for your job
Subscriptions to professional journals
Telegrams
Telephone calls
Tips
Travel expenses while away from home
Typewriter
Vehicle expenses

Of course these deductions are valid only if they are genuine business expenses, and other deductions may be allowed.

Incidentally, you cannot deduct expenses for meals during regular or extra work hours, the expenses of going to and from work, or education needed to meet minimum requirements for your job or to qualify you for a new occupation (as opposed to education required to keep your job).

The way you file these expenses has changed in recent years, and it seems to be getting harder and harder for a freelancer employed by an agency to deduct much. There are two parts to this puzzle. The first part is your business expenses and the second part is the amount reimbursed, if any, by your agency.

If an agency reimburses you, perhaps in the form of a per diem, try to make sure that the reimbursement does not appear on your W-2 form. You can do this by adequately accounting to your agency for your expenses, and making sure that the reimbursement does not exceed the expenses. Of course this requires the agency's cooperation, and is more work for them, but if you don't do this the agency is supposed to report the reimbursement on Form W-2 as part of your income.

In previous years this didn't matter too much. You would fill out Form 2106 showing that you had expenses up to or above the reimbursement, and then deduct the reimbursement off your taxable income (in other words, the agency told the IRS that it was income, and then you showed the IRS that it was a reimbursement so the IRS let you take the sum off your income). But the law has changed. Once that reimbursement appears on your W-2, the IRS regards it as income and it cannot be removed. Your only option is to deduct the expenses on Schedule A, the Itemized Deductions form (there are certain expenses that can be deducted directly from your income, but they are not usually applicable to technical writers; for example, the expenses of a performing artist may be deducted in this way, as may certain work-related expenses of a disabled person).

This is the procedure for deducting expenses. First fill out Form 2106. This shows your expenses and how much the agency reimbursed you, including any sum the agency listed on Form W-2 as income. You will deduct the reimbursements from your expenses, leaving the unreimbursed amount, or the excess reimbursement.

If you have an excess reimbursement, more money paid to you than you spent in expenses, you must report the excess reimbursement on Form 1040. The reimbursement in excess of the expenses becomes taxable income. If, on the other hand, your employer reimbursed you less than your expenses (or if you received no reimbursement), claim the excess expenses on Schedule A.

Now we arrive at the catch. Having kept all the records and messed with Form 2106, you may find you can't deduct the unreimbursed expenses anyway. They are deducted as Miscellaneous Deductions on Schedule A (Itemized Deductions). Unfortunately many people can't use Schedule A, especially people who don't own a home. All taxpayers have the option of either filing Schedule A or taking a standard deduction. For example, a married couple filing jointly can either deduct $5,700 from their adjusted gross income or can file Schedule A. Of course, if their itemized deductions are less than $5,700 there is no point using Schedule A, and unless they own a home, give a lot of money to charity, or lose a lot of property to fire or theft, they are unlikely to exceed the $5,700.

But even if they do file Schedule A, only the amount of the Miscellaneous Deductions more than 2 percent of their adjusted gross income is deductible. For instance, if they have an adjusted gross income of $50,000 and $1,500 of Miscellaneous Deductions, only $500 will be deductible (2 percent of $50,000 is $1,000; $1,500 minus $1,000 is $500). Depending on where on the tax table their other deductions put them, the $1,500 in expenses may end up saving only $75 to $140 in taxes.

Moving Expenses

Although many of the expenses of living away from home may not be deductible (because, as I've explained, the IRS may regard you as being in your "tax home"), some of those expenses may be deductible as moving expenses. You can deduct these expenses only if you itemize on Schedule A, but they are not limited by the 2 percent floor—the entire sum can be deducted. You have to use form 3903 to report the moving expenses, and then deduct the sum on Schedule A.

There are a couple of rules regarding moving expenses. First, the distance from your new job to your old home must be more than thirty-five miles greater than the distance from your old job to your old home. Got that? In other words, if you stayed in your old home you would add thirty-five miles or more to your one-way commute by taking the new job. Second, you must stay in the new area and have full-time employment for at least thirty-nine of the next fifty-two weeks. So if you move to area A for a contract, and then move to area B after six months, you won't be able to deduct the moving expenses. However, there are various waivers; if you lose your job for any reason other than willful misconduct or resignation, the thirty-nine-week limit can be waived, but I don't know if that would be applied to a contractor in a limited-duration contract.

So, what exactly can you deduct? The direct expenses of moving are fully deductible (travel and lodging while en route) and the cost of moving your personal effects and household goods. You also may deduct other expenses up to certain limits: the cost of pre-move house-hunting trips, temporary living quarters for thirty days, and the cost of selling, buying, or leasing accommodation.

If you are unable to deduct living expenses as business travel expenses (or don't want to risk an audit), look at the possibility of using moving expenses instead. Since they are not subject to the 2 percent floor, they may be more profitable anyway.

The Independent Freelancer

The independent freelancer, called a *sole proprietor* by the IRS, has a few advantages and one big disadvantage, as far as taxes go. The independent can deduct expenses directly off his income, before paying FICA or income tax. Because the expenses are coming directly off your income you don't need to use Schedule A, so you are not caught in the "2 percent of adjusted income" trap—whereas the employee of an agency *may* be able to deduct *some* of his expenses, but will always pay FICA first. The big disadvantage for the independent, though, is that he will pay much more FICA and Medicare insurance. Employees pay 7.65 percent FICA on the first $53,400, and 1.45 percent Medicare insurance on the next $71,600. The independent freelancer pays 15.3 percent and 2.9 percent, respectively, for an extra $5,123.30 a year (if he makes $125,000, that is). It's not quite that bad, actually, because this extra amount can then be deducted off his taxable income, so the extra FICA and Medicare insurance that he pays ends up about $3,535. Also, starting in 1991, self-employed people calculate the income on which FICA must be paid by deducting 7.65 percent from their self-employment income. So, if you earn $47,000, you pay self-employment tax on $43,404.50.

Independent freelancers file taxes in a different way. Because no one is withholding taxes, you have to do it yourself. The IRS requires you to estimate your taxes for the year and then pay an equal sum every three months.

(In the past the IRS required independents to pay installments in the middle of April, July, October, and January. However, in recent years the October payment was pulled back to September. This was probably due to what *Newsweek* called "stupid budget tricks." Since the government's financial year begins on October 1, Congress realized that by bringing the October estimated tax payment forward one month they could make the current year's budget look better than it actually is. Of course they have to find another stupid trick to replace the money lost to the following year, so the third "quarter" payment will continue to be paid in September each year. Incidentally, you can forget the estimated tax payment for the middle of January if you file your final tax return and pay all tax owed by the end of January.)

You are now (1991) required to pay 90 percent of the tax you owe by the middle of January—with at least 22.5 percent (25 percent × 90 percent) paid by each installment date—and the rest when you file your tax forms (between January 1 and the middle of April). There are a few loopholes, though. The most commonly used is this: Make sure that by the 15th of January you have paid the same amount you paid in the previous year, even if it is far less than 90 percent of the tax you will owe for the current year. Then pay the rest by the middle of April.

Don't worry about getting your estimated income exactly right at the beginning of a year (none of us knows for sure how much we will make during the year). Just reestimate during the year, and amend your estimated payments correspondingly. Estimated tax payments are filed on Form 1040-ES, by the way.

When you file your final tax return you will have to fill in these forms: Form 1040, Schedule C (Profit or Loss from Business), Schedule SE (Self-Employment Tax), and Form

4562 (Depreciation and Amortization) if you bought an expensive piece of equipment for your business.

SCHEDULE C

Unlike freelancers working as agency employees, an independent contractor can deduct all valid expenses on Schedule C. These are the major categories:

Advertising	Legal and professional services
Bank service charges	Office expenses
Car and truck expenses	Rent
Commissions	Repairs
Depreciation from Form 4562	Supplies
Dues and publications	Taxes
Freight	Travel, meals, and entertainment
Insurance	Utilities and telephone
Laundry and cleaning	Other Expenses

All the expenses listed earlier in this chapter will fit in one of these categories. Also, 25 percent of your medical insurance premiums may be deducted straight off your gross income on Form 1040 (the rest may be included with your itemized medical deductions on Schedule A).

Use the cash method of accounting, by the way. With the cash method you report income in the year you receive the money, and expenses in the year you spend the money. With the Accrual method you report income in the year that you earn it, regardless of when you get the money, and expenses when you incur them, even if you don't pay till the next year. The accrual method is a much more complicated method with more stringent reporting and accounting requirements, and is unnecessary for freelancers.

Some of Schedule C is not applicable to your business. For example, the "cost of goods sold" is used by businesses that manufacture or purchase goods for sale, so few freelancers will need to worry about them. Otherwise the form is quite simple. In Part I you list your gross income. In Part II you list all your business expenses and then subtract the expenses from your income to arrive at a profit or loss. You then report that figure on Form 1040.

FORM 4562: DEPRECIATION AND AMORTIZATION

If you have purchased some kind of asset that you use in your business, you must complete Form 4562. An asset is an item that is expected to have a life of more than one year: vehicles, computers, tools, telephone equipment, and so on. Of course, if you buy an inexpensive tool like a screwdriver, you don't need to fill in this form (just claim the expense under "Supplies" on Schedule C). Use Form 4562 for goods costing hundreds or thousands of dollars.

You can choose to depreciate the item, or deduct the first $10,000 using the Section 179 deduction. If you choose to deduct the cost, you must still fill in Form 4562, but then deduct the expense on Schedule C. If you want to depreciate the expense you must decide what type of equipment it is and over how many years it will be depreciated. "Depreciation" simply means deducting the expense over several years.

For example, you buy a computer for $8,000. You can use the Section 179 deduction and claim the entire sum on Schedule C—reducing your taxable income by $8,000 that year— or you may decide to deduct a portion of the cost in each of the next five years. Refer to IRS publication 534 for more information.

If you have a computer I recommend that you get a tax program. Get one that allows you to do "what if" experiments (most do), so you can plug in different numbers and see what effect they will have on your taxes.

Forms You Need

Here is a list of some of the tax forms you may need.

Form	Title
1040-ES	Estimated Tax for Individuals
1040	U.S. Individual Income Tax Return
1040A	U.S. Individual Income Tax Return (an easier form than 1040, but you can use it only if your income is less than $50,000 and you don't use Schedule A)
1040EZ	Income Tax Return for Single Filers with No Dependents
2106	Statement of Employee Business Expenses
3903	Moving Expense Adjustment
4562	Depreciation and Amortization
4782	Employee Moving Expense Information
8829	Expenses for Business Use of Your Home
Schedule A	Itemized Deductions
Schedule C	Profit or Loss from Business or Profession
Schedule SE	Computation of Social Security and Self-Employment Tax

Tax Books to Refer to

Get hold of a good tax book, such as *J. K. Lasser's Your Income Tax* or *The Consumer Reports Books Guide to Income Tax Preparation*. These will answer many of your questions. (Many of the tax books seem little clearer than the IRS's own publications, but they have the great advantage of putting all the information you are likely to need in one place.) You also can get the following publications from the IRS:

Publication	Title
17	Your Federal Income Tax
334	Tax Guide for Small Business
463	Travel, Entertainment, and Gift Expenses
505	Tax Withholding and Estimated Taxes
508	Educational Expenses
521	Moving Expenses
529	Miscellaneous Deductions
533	Self-Employment Tax
534	Depreciation
535	Business Expenses
560	Self-Employed Retirement Plans
583	Information for Business Taxpayers
587	Business Use of Your Home
589	Tax Information on S Corporations
590	Individual Retirement Arrangements (IRAs)
917	Business Use of a Car
936	Limits on Mortgage Interest Deduction

You can write or call the IRS and have them send you these booklets and the forms you need. (See IRS, Appendix F.) Or you can go to your public library and refer to the books and forms there, and make copies of the parts you need.

This chapter may have convinced you to do your own taxes, or to find a good accountant. You've got enough problems without this tax nonsense, and you may not want to spend the time and energy learning the tax law, but a few evenings spent reading (curled up in front of the fire?) will teach you all you need to know, and after the first year it is just a matter of keeping up with a few small changes.

Incorporation

Most contractors are probably better off being sole proprietors (self-employed) than employees of their own corporations. Incorporating causes more paperwork and more tax: franchise tax, payroll taxes, corporate income tax, and then personal tax on the salary and dividends your corporation pays you. There are some advantages: Medical insurance could be deductible as an employee benefit, for example, and you could buy up to $50,000 life insurance coverage and charge it to the corporation. Corporations also limit your legal liability in some situations.

A special form of corporation exists for small businesses, the Subchapter S corporation. The S corporation allows you to avoid corporate income tax. If you do decide to incorporate, spend some time investigating the pros and cons; talk to a tax accountant or lawyer, or do a lot of reading. You can save money by incorporating for yourself. There are plenty of books and even computer programs that will show you how to do it (no, you don't need a lawyer), but at least make sure you understand what you are getting yourself into.

21
Are You Really an Independent Contractor? What the IRS Has to Say

The IRS seems not to like independent contractors. Too many contractors have avoided paying taxes in the past, and too many companies have avoided paying their share of taxes by calling employees contractors. Furthermore, if a contractor defaults on taxes, the IRS may have trouble getting the money, although if a large company defaults, the IRS probably will have more luck. It's easier for the IRS to keep an eye on a few thousand companies rather than hundreds of thousands of contractors. An independent contractor is more likely to figure he can "slip by" without paying than is a company that the IRS can easily find.

So the IRS has a list of twenty factors it uses to determine if you are a true independent contractor, to find out if there is anything that would distinguish you from your client's regular employees. If you *are* an independent then your client can continue paying you all your money, without withholding any taxes. You are then responsible for paying those taxes. If you are *not*, the client has to withhold taxes, and the IRS may make them pay back taxes that they should have withheld earlier, and penalties. In addition, you may lose your self-employment income pension plan.

Here are the twenty factors the IRS examines:

- Instructions
- Training
- Integration
- Services rendered personally
- Hiring, supervising, and paying assistants
- Continuing relationship
- Set hours of work
- Full-time required
- Doing work on employer's premises
- Order or sequence of work set by employer
- Oral or written reports
- Payment by the hour, week, or month
- Payment of business and travelling expenses
- Furnishing tools and materials
- Significant investment by the employee
- Realization of profit or loss by the employee
- Working for more than one firm at a time
- Making services available to general public
- Right to discharge
- Right to terminate

You can find these twenty factors in Revenue Ruling 87-41. I have explained them below, in the form of a questionnaire. A Yes answer to any of the first fourteen questions takes you one step further from independent contractor status.

1. **Instructions:** Are your required to comply with another person's instructions about how, when, and where you work?
2. **Training:** Did the client require that you take a training course to learn how the client wants you to do the job?

3. **Integration:** Are your services integrated into the client's business operations?
4. **Continuing Relationship:** Do you have a continuing relationship with the client (even if you work irregularly but frequently)?
5. **Set Hours of Work:** Does the client set your hours?
6. **Full-time Required:** Does the client expect you to work full-time, effectively restricting you from working for other clients?
7. **Doing Work on Employer's Premises:** Do you work at a location specified by the client?
8. **Order or Sequence of Work Set by Employer:** Does the client have the right to establish the routines and schedules that you follow in your job?
9. **Oral or Written Reports:** Does the client require you to submit regular oral or written reports?
10. **Payment by Hour, Week, or Month:** Does the client pay you for the time worked, rather than paying a set fee for the job?
11. **Payment of Business and Travelling Expenses:** Does your client pay your business or travel expenses?
12. **Furnishing Tools and Materials:** Does the client furnish your tools and materials?
13. **Right to Discharge:** Does the client have the right to fire you?
14. **Right to Terminate:** Do you have the right to quit at any time without incurring liability?

A Yes answer to any of the next six questions helps your case, making your status appear closer to that of a true independent contractor.

15. **Services Rendered Personally:** Do you employ someone else to do some of the work covered by the contract with the client? (You may have your own employees, or use subcontractors.)
16. **Hiring, Supervising, and Paying Assistants:** Do you hire, supervise, and pay assistants according to a contract in which you are responsible for providing materials and labor?
17. **Significant Investment:** Have you invested in facilities and tools that are not normally maintained by an employee?
18. **Realization of Profit or Loss:** Do you have a possibility of economic loss through an investment in materials, equipment, or salaries?
19. **More Than One Client at a Time:** Do you work for more than one client at a time?
20. **Making Service Available to General Public:** Do you regularly and consistently make your services available to the general public? Do you advertise or market your services?

Incidentally, it doesn't matter what your client calls your position or status—the contract may state that you are an independent contractor, but if the IRS decides that the facts show otherwise, the contract is irrelevant.

Now, how does all this affect you? If the IRS decides you are an employee, it will most likely ask the client, not you, for back taxes. The client also may have to pay the same sum as a penalty. The client also will have to treat you as an employee, and some clients may decide to get rid of you instead (although as your employer they may have to show good cause, depending on the state you work in). Furthermore, if the IRS decides you are an employee, you are not eligible to have a self-employed person's pension plan (an individual 401(k), SEP, or Keogh plan), and lose your Schedule C (Business Profit and Loss) deductions.

With the IRS campaigning against companies using contractors, companies are going to start protecting themselves. And how can they protect themselves? By hiring through technical service agencies, of course! The Tax Reform Act of 1986 allowed companies to treat contractors as nonemployees if they were hired through a technical agency that employed the contractor. This means that as more companies get audited (and fined or assessed back taxes), fewer independent contract positions will be available.

There is a "safe harbor," though. Section 530 of the Revenue Act of 1978 provides protection cases where the IRS decides that the company was incorrectly treating an employee as a contractor. If the company meets certain criteria, the IRS cannot assess penalties or back taxes. First, the employer must have consistently treated the contractor as such; the firm must not have changed the contractor's status from that of an employee. It must have paid this person as a contractor, and it must have given the contractor work that differs in some way from that of its employees.

Second, the company must have filed the correct tax forms, in particular Form 1099NEC (Non-Employee Compensation). And third, the company must have had a reasonable basis for treating the contractor as such; perhaps an IRS precedent treated others in that profession as contractors, or maybe it is a common practice in that industry. Incidentally, an amendment to Section 530, Section 530(d), makes this "safe harbor" applicable only for independent contractors under contract to the client they are working for, not to those with technical service agency contracts. So if a company hires an agency to provide a contractor, and the agency pays the contractor on a 1099, as an independent, the safe harbor is no longer available, to either the agency or the client. Thus, since the IRS may find the client liable—in addition to or instead of the agency—companies can find themselves at more risk than if they had hired an independent. If you wish to read more on the subject see "Determining Employee or Independent Contractor Status" in *The Tax Advisor* of October 1989. Your library may have a copy.

An agency I know uses these IRS problems as a marketing tool. They visit a company and show them an information package describing various laws and their effect on companies. The agency says, "We can keep the IRS and the Department of Labor off your back, if you hire our people."

Industry insiders say that the 1990s will see the IRS move in a lot more employee/contractor cases, especially in California. The IRS is looking for test cases and companies to make examples of, so firms returning a lot of 1099 forms to the IRS (the form declaring payments to contractors) will be targeted for audits. Things are unlikely to change in a hurry, though. The IRS probably will go after companies using a lot of low-paid contractors first, such as construction firms.

There is also a rumor in contractor circles of a new law or ruling that would state that any client paying a contractor more than 51 percent of his income would be regarded as the contractor's employer; in effect he would have an employee/employer relationship with that client, and a contractor/client relationship with the other clients. I can't imagine how on earth this could work, however. How is a contractor (or client) going to know on January 1st which company will have paid 51 percent of his income by the end of the year? And what about a contractor who spends 60 percent of his time working for a company, but only receives 40 percent of his income from that company? However the law changes, don't worry; it's not the end of the road for independents. There are a number of things you would be able to do to protect yourself.

First, there will still be some contract jobs that are quite clearly and unambiguously independent contractor or consultant jobs—contracts in which you work at home and use your own computer equipment, for example, as I do. If you set your own hours and are not supervised or controlled in any manner, for example, you will probably be okay.

Second, some clients will accept contractors, but put them on their payroll as temporary employees; the company withholds all the taxes mandated by law, but doesn't provide any of the usual employee benefits. I know of one company that does this—the contractor makes his deal with the department manager, and then the accounting department adds

10 percent to that rate and withholds taxes. Also, some clients will still be willing to take the chance or will remain ignorant of the law, especially small companies that don't use many contractors and so don't attract much attention.

If you find a contract like this, do everything you can to get as close to independent contractor status as possible: Avoid having to provide the client with status reports; arrange a "flexi-time" schedule that allows you to set your own hours; try to do some of the work at home; regularly solicit work from other companies; and try to do several contracts each year, even if you have one main client and several other small contracts. Avoid working for one client for an extended time (of course this is a double-edged sword; you don't want to be seen as a contract hopper, nor do you want to leave long, profitable contracts). Set up a home-office, and use it as a business place. Get a business name, business cards, and even a brochure, and mail them out to potential clients.

Don't make it difficult to hire you, though. I know freelancers who tell clients they *must* be allowed to work flexi-time and to work a significant part of the time at home, and that the client mustn't expect progress reports, and so on. This is fine in some circumstances, but most contract jobs are with clients who require someone in the office, working with their equipment and employees, during regular business hours. Don't scare off these clients.

Third, you may be able to become a true *consultant*, moving on to Step Three. By *consultant* I mean someone who is more than simply a temporary employee, someone who comes to the company and provides specific skills for a specific project. The consultant typically charges a set fee of some kind, and may have his own staff, or subcontract for the project. I discuss consulting in more detail in chapter 22.

Fourth, if the independent contract market gets tough, you might consider starting your own technical service agency. Find a few independent contractors and a good accountant, and incorporate. You can then use the company as a vehicle to sell your services. The client would pay your company, rather than pay you on a 1099, and you would arrange for the account to be paid jointly by all the contractors (the owners). All the money could pass through the company in the form of salaries, with each contractor getting what his clients paid to the corporation, minus expenses. Another advantage to setting up such a company would be the benefits you could provide, in particular medical insurance. Your company may be able to get a much more favorable policy than you could get independently. Of course, a problem would arise when one of you wanted to buy capital goods, such as a computer; if the individual bought it, it would be only partly deductible from his taxes, but if the corporation bought it, it would belong to everyone. Some writers have already begun setting up such agencies.

Another alternative would seem to be setting up your own corporation. Find an accountant, or incorporate yourself using one of the many books or computer programs, perhaps using a Subchapter S corporation. Unfortunately this may not work. According to *The Tax Advisor,* an amendment to the Tax Reform Act of 1986 stops contractors from forming Personal Service Corporations in order to bypass the effect of Section 530(d). If you want to set up your own corporation, find an attorney who is familiar with this aspect of the law.

Fifth, if you don't want to incorporate yourself, you might be able to find another freelancer who has a small agency and is willing to take on your contract and pass all but a small amount back to you. I wouldn't advise you do this unless the freelancer has an accountant and is willing to do all the taxes and paperwork, rather than pay you on a 1099—after all, you may as well handle the contract yourself if you are going to be paid as an independent. Calculate exactly what it will cost the freelancer to do this (in accountant's fees and payroll taxes), and then add a dollar or two profit for him. If he doesn't think that is enough, remind him that he won't have to do anything for the money, that you are paying the accountant.

I have a programmer friend who is working through a "front" corporation like this, but they pay him on a 1099. The client thinks it is free of all responsibility, but the client is wrong. The IRS recognizes "joint employment," so if they assess the agency with back taxes or overtime pay, and the agency cannot pay, the client could be dragged in.

Finally, the most important thing you can do is *to pay your taxes promptly and correctly.* Many contractors invite the wrath of the IRS by "forgetting" to pay taxes (for five or six years) or claiming ridiculous deductions. Once the IRS starts auditing you they may decide to question your employment status. Audits are annoying, expensive things, even if you "win" (I know, I've won one), so do your best to avoid them. If you lose your audit you may lose the right to your pension plan, and bring the IRS down on your client. If you don't pay your taxes properly you are doing both yourself and your client a disservice.

Don't let all this put you off seeking independent contracts. I've included this information because you need to be aware of it—it can help you stay on the right side of the IRS—and because it may become more critical soon. But at the time of writing there are still many independent contracts available. Many companies have stopped using independents, but many haven't, and I don't know of anyone who has been audited for employment status. So go looking for those contracts. They are well worth it.

V.
The Third Step in Freelancing

22
Where Next?

That's a big question: What do you intend to do next, what is the *third* step? Once you have managed to make a good living selling your services on contract, what should you do? Well, many people spend the rest of their careers doing more or less the same—selling their services for 50 percent, 100 percent, or 150 percent more than they could make as employees. That is not a bad way to go. With financial discipline you can do a better job of providing for your family and your retirement than you could if you were a captive employee.

This book has explained how to work like an employee but get paid like a consultant. Many contractors like to use the word *consultant* when people ask them what they do, but they are misusing the word. A contractor fills a job slot temporarily, perhaps for a week or two or maybe a year or two. He often does the same work as an employee; in fact in many situations you will work side by side with the client's employees, doing the same work in the same conditions. Contractors remain cogs in the machine, but become well-paid cogs.

A true consultant, on the other hand, is in a different position. The consultant may have had to create the position for himself, using marketing skills to convince the client he should hire the consultant to deal with a special problem, a problem the client's company does not have the resources to deal with. The consultant works independently, in his own way, and is not controlled or supervised by the client. The consultant may even hire other people to work on the project, and supervise them. Also, the consultant usually receives a set fee, because the consultant is selling the solution to a problem or a finished product, not his time.

Some contractors may wish to continue down the road to consulting. Consulting generally requires much more marketing, and more sophisticated marketing, than mere freelancing. It also may require more experience and ability. Consultants traditionally charge much more than contractors, but may also bill for less time, having to work more at selling themselves and in administrative tasks.

Is it worth it? Successful technical writing consultants can do very well, making $100,000 a year or more. But those who *don't* sell themselves well are often no better off than their contracting cousins—who may be working as cogs in the machine but are also free of much of the hassle involved in consulting.

In *The Complete Guide to Consulting Success* Howard Shenson provides a table showing the average pre-tax income of consultants in a variety of disciplines. For example, the average engineering consultant makes $96,311 while the average data processing consultant makes $81,008. Shenson believes the average income of all consultants to be about $91,102. Healthy though these incomes may be, they are within reach of many technical writers who are contracting rather than consulting. A writer in New York, for example, making $50 or $60 an hour and working 1,800 hours a year will make $90,000 to $108,000.

But you may be suited to consulting, and you may have the skills that will allow you to break the $100,000 barrier. What then? Consulting is a completely different way of working, so you should continue your education by reading books on consulting. An excellent primer is Shenson's book, mentioned above. (See "Other Material," Appendix E.) It has a one-year money-back guarantee, and seems to be more informative than most bookstore books on the subject.

What do I mean by consulting? Generally, I'm assuming that a company gives you a project and you complete it. The company doesn't watch over your shoulder, they trust you to provide a document of an acceptable standard in the time allotted. You are not a contract worker, paid by the hour to do whatever job needs doing that day. In fact, you are usually working at home, using your own computer (though in some cases the company may provide you with the equipment). You may charge by the hour, but you are more likely to charge by the project, a set fee for the entire project regardless of the time it takes you.

In order to charge by the project, you must be able to estimate how long a job will take. I learned how to do that by logging my hours while working on my books. I used a stop watch, turning it on when I sat down to work, and off when I stopped—whether to take a break, answer the phone, or go to the bathroom. Try this and after a while you will find out how quickly you work, and be able to estimate how long a project will take. I don't mean that you will be able to say, "This book will take 320 hours." What you will be able to say is "This book will probably take 320 hours, but if it turns out to be a lot more complicated than I imagine, it could take as much as 400 hours." That's a big range, but it doesn't necessarily matter, because you can charge a much higher set fee than hourly fee, *if* you work quickly.

For example, say you estimate the book to be a 300-page book, and you feel that in your area $100 a page would be a reasonable fee. You propose a fee of $30,000. Now, you think it will take you 320 hours; if it does, you have made $93.75 an hour. If your worst-case estimate is right, though—400 hours—you have made $75 an hour. Certainly you want to make the higher hourly rate, but so long as the lower rate is still acceptable, your worst-case estimate won't hurt too much if it comes true. (What if it takes *600* hours? If it does, you haven't yet learned how to estimate your time!)

What do you base your fee on? I know a lot of writers who base their fees on what they want to make per hour. That's the wrong way, because what they want to make is usually based on the hourly fees they are used to making as contractors. One consultant, for example, told me that she estimates how long the job will take and then multiplies by $35 an hour. Your fee should be based on "what the market will bear." Consider how much the company would pay if they hired somebody from a technical service agency. For example, in Dallas a company would usually pay somewhere from $37 to $45 an hour for an experienced technical writer. Multiply the *top* number by the number of hours it would take to do the job—not the number of hours it would take *you*, because you have estimated *real* hours, not company-work hours. No, you need to figure out how many hours a normal contractor would take to complete the project, working eight to five. Now, eight to five with an hour for lunch is *not* eight hours. Subtract the time spent on breaks, getting back late from lunch, talking to co-workers, unproductive company meetings, tracking down supplies, on the phone, and so on, and an eight-hour day is actually about four hours of work. (I'm not exaggerating: Take a stopwatch to work with you and time the hours you and your colleagues spend in *productive* work—you might be surprised.) But the client pays for both the productive and unproductive time.

Now, if you estimate that a job takes 320 hours of *productive* work, multiply the hourly rate—in this case, $45—by 640. That 320-hour project could easily cost the company $28,800, yet the consultant who multiplies her hours by $35 would bill only $11,200. Of course there's a catch; as you can see, you can make a lot of money by completing a job more quickly than most other writers, while charging the same. This means that you will work harder than you would normally, of course (but after all, nothing's for free!).

Not only can you make more money by working more quickly, but you can increase your income by becoming the sort of writer whom companies are willing to spend more money on. The 640-hour estimate we just made assumed that the job would be completed satisfactorily by a contractor working for 640 hours. In many cases contract (and full-time) technical writers work for months or years and produce little or nothing. I know writers who jump from job to job making good money, without doing anything worthwhile for their clients or employers. Many companies have lost tens or even hundreds of thousands of dollars by hiring bad technical writers. You can improve your consulting career by proving that *you* can be trusted to deliver the goods, on time and up to spec. While some companies operate in the "warm body" mode—they hire the cheapest writer they can find—many actively seek writers with good reputations and are willing to pay more rather than to take a risk with an unknown commodity.

So how can you build your reputation? The first step, of course, is to do a good job for all of your clients. The next step is to build your network, let people know who you are and what you can do. But you can do more. Write a few magazine articles, publish a book or two. Write a column for a local paper or computer magazine. You may not make much money—although if you do the right computer book you could make quite a bit—but having published work will help you build your image as a *professional* writer.

How can you find the high-paying consulting projects? Well, if you have followed the techniques in this book, you will already be meeting with people who can hire you as a consultant. My first "set fee" project was with a company that originally wanted to hire me on an hourly rate. I told them I would rather do the work at home, and convinced them that the benefit of knowing from the start how much the work would cost outweighed the disadvantages, if any, of having me work at home. Not all projects can be done away from the customer's site; some require specific equipment that is only available there. But many projects *can* be done at home, and the first step to getting this work is to recognize the opportunity and propose that you work at home. Remember, nevertheless, you are selling something and you must make it sound as though the customer will *benefit*. And proposing a set fee is a good way to do so.

If you become known throughout your network as a writer who works for a set fee, you may get offers out of the blue from companies that want that sort of relationship. Most writers *don't* want to work on a set fee—they are scared of underestimating—so if you are known to work like that, you will sometimes get good referrals.

How else will you find consulting work? Read Howard Shenson's book, and Robert Bly's *Secrets of a Freelance Writer*. They contain many good ideas. One method I can vouch for is writing magazine articles. After one of my articles appeared in *Software Maintenance News* I received two offers of work, one from a local company and another from a company in New York (both were at-home, set-fee projects). I have already begun working on the local project, and I'm currently negotiating with the New York company.

Another method I can vouch for is direct mail. A mailing of 1,000 pieces to computer software companies brought me a 2.6 percent response rate (which is pretty good for direct mail). One respondent asked me to write a magazine article (the *Software Maintenance News* article that brought me work), one asked me to begin work on a 400-page manual immediately, and one asked if I could write a computer manual for them. (I *didn't* take either of the big projects, though, because by the time I received the offers I was swamped with work). The other responses ranged from "We may be interested later" to "We've got something soon," though admittedly none of them have turned into anything so far.

Don't just mail your résumé to 5,000 companies, however; you may get a few job offers, but you won't get consulting work. Before you produce a direct-mail piece, do some research. In particular read Jeffrey Lant's *Cash Copy*. You have to give your target companies a *reason* to respond, not just say, "Here I am, come and get me." There's an art to it, so before you invest a lot of money in mailings, make sure you can produce a reasonable mailing piece. My mailing cost about $730 for 1,000 pieces, by the way, though mailing rates have gone up since then. You can buy mailing lists, or build your own from directories you can find in your local library, or by compiling lists of advertisers in magazines.

How else might you sell your work? Contact consulting programmers and begin a professional relationship; for every program produced there is a manual waiting to be written. Talk to local printers and find out if they can recommend you to companies that are printing manuals. Try advertising in industry-specific publications. (Shenson says this is not an efficient way to find business, so I would call other consultants who advertise in publications and go ahead only if they had a good response.) You might try looking for work using the CompuServe computer network, teach a technical writing course, produce a newsletter for local companies, or give a freelancing seminar at your local STC chapter. Shenson claims the most productive techniques are using references from previous clients, giving seminars, giving speeches, and publishing newsletters.

Finding consulting work is another book's worth of material, so if that's what you want to do, read Bly, Shenson, and some of the other authors in Appendix E.

You might want to try the computer-book field. A few of the writers who entered the field in the early days of the IBM PC—or a little before—are now millionaires, selling literally millions of copies of their books. That's harder to do now, as more authors and more publishers have jumped in and fragmented the market. However, it is still possible to make a good living. Matt Wagner of the Waterside Productions literary agency told me it has 200 technical writers on its roll. "A good writer entering the field who is prepared to work hard can make $60,000 a year," he told me; "we have lots making $100,000." Advances range from about $3,000 to $20,000, and royalties are commonly around 10 or 12 percent of the net receipts (about 5 or 6 percent of the bookstore price). Of course most of Waterside's writers don't get rich writing computer books: "Twenty percent of our writers make 80 percent of the money," Wagner told me, and many of Waterside's writers make $20,000 or less each year—though these writers usually also have full-time jobs.

Maybe you don't want to become a consultant or write computer books; perhaps you have something else in mind. Contracting is especially useful for people who want to pursue other goals, people who need free time or available money to pursue other avenues of advancement. Technical writing is perfect for the freelance writer, as I discovered. Working contract allows you to work when you want, and to build enough savings to stop work for long periods. Contracting allowed me to stop work for two months one summer to finish my first computer book (the *Illustrated Enable/OA* for Wordware Publishing), a large project that would have taken me at least one year (and incredible self-discipline) to complete if I had been writing only in the evenings and on weekends. That book will never make much money, but it was a stepping-stone to more lucrative books and better-paying consulting work.

I have also written two computer books for SYBEX, and I'm working on a third. The money and flexibility provided by freelancing is allowing me to write and publish books much more quickly than I could otherwise do. I hope that in a couple of years I can leave contracting entirely and write books full-time. While these are technical books, you could use the same time and money advantages for nontechnical projects.

Maybe you want to write a novel, a book on CIA influence on Latin American elections, or a biography of Pope Boniface VIII. Contract technical writing can provide the time and money that will let you do that. Had I been a captive employee limited to a couple of weeks' vacation a year, and a salary 40 percent or 50 percent of what I now earn, I don't know if I would have had the discipline or energy to start a writing career. It is not easy to come home at night to write for an hour or two, especially when you are working on a 400- or 500-hour project and the work seems to stretch endlessly before you.

You may have other goals that freelancing can help you reach. Perhaps you have only a few years left to retirement, and would like to increase your nest egg. Maybe you have decided to change careers—why not work as a freelancer while you go back to college, or do it for a year or two to save enough to go back to full-time work? Perhaps you have always wanted to travel around the world, but could never afford it. Contracting for a year or two might allow you to see your dream come true. Or maybe you've always wanted to have your own business, to move back to your home town and open a restaurant or small hotel. A few years' contracting and investing could help you achieve just that.

Most technical writers can substantially increase their incomes by freelancing, and some can literally double or triple what they earn. You can use that extra money to make life materially more comfortable in the short run—buy a fancy car, put in a swimming pool, buy jewelry—and go on living life much the same as before. But you also can use the money to change completely the way you live your life, to get out of any ruts you happen to be stuck in, or to do something you always had the ambition to do but for which you never had the time or money.

Appendixes

Note: Costs cited in the appendixes that follow are accurate at the
time of publication and are only meant as relative indicators.
Please check for the most up-to-date quotes.

Appendix A
Contractors' Publications

There are several publications targeted at contractors. Two in particular are of special help to *road-shoppers*, contractors who travel around the country working on different contracts—six months in New York, a year in North Carolina, eighteen months in California, and so on. These magazines contain agency ads and listings of contracts, and several other services that may be useful. One of these magazines, *PD News*, also has some services that could be useful to any freelancer, even those who never leave their home towns.

Remember, though, that many of the listings may not be "real," because some agencies use the listings and ads as a way to build up their résumé database. Also, as an agency employee pointed out to me, some of the agencies don't even use the résumés that they receive from responses to the listings, unless the respondent lives in the agency's area—these agencies use the magazines to build their database of *local* résumés.

Still, many people use these magazines to successfully road-shop, and so can you. ("I have been lucratively employed for the past nine years; this is a direct result of your efforts," one job-shopper wrote to *PD News*.) Others just use the magazines to keep in touch with what is going on in contracting and to use their special services. If you road-shop for more than a year or two, by the way, you will want to build your own list of agencies as you go—you could collect the names and addresses of all the agencies in each town in which you work. One contractor told me he had a computer database listing 264 agencies around the country.

PD News

P.O. Box 399
Cedar Park, TX 78613-9987
(512) 250-8127
FAX (512) 331-6779

The nation's first publication serving the contract employment industry. A weekly publication with display ads and listings of contracts throughout the nation. Unlike *C.E. Weekly*, *PD News* also includes articles by "job-shoppers" about taxes, overtime rates, finding work, working in different states, and so on. *C.E. Weekly* usually limits itself to letters from readers. I used to write regularly for *PD News* (only occasionally now), so if you subscribe, tell 'em I sent you. 36 pages, about 175 agencies per week. $51 for 52 issues, $6 for one issue. The subscription includes the annual *Directory of Technical Service Firms*, listing about 800 agencies. Nonsubscribers can purchase the directory for $10. (Subscribers can buy the directory on computer disk for $10.)

PD News also offers PD Passport, a special service for freelancers. You get the following benefits for $89 a year:

- *PD News* subscription
- The annual *Directory of Technical Service Firms*
- Free Hot List listing (see description below)
- Free acknowledgment postcards for your letters to agencies
- Free mailing labels with agency addresses
- Membership in NASE, the National Association for the Self-Employed, including a subscription to *Self-Employed America*

By associating with NASE, PD Passport is also able to offer the following services:

- Free Care-by-Air National Ambulance Service membership
- Free $2,500 heart attack insurance and up to $10,000 accidental death coverage
- Medical, dental, and disability insurance at group rates
- Credit union membership
- Paine Webber pension plan, and a high-yield annuity plan

- A 15 percent discount on Tax Audit Answers services
- Free business advice from an NASE small business consultant
- Free personal business and health publications
- Discounts on airfares (up to 7 percent) and on special vacation packages

Your NASE membership also brings savings on: car rental, K-Mart automotive service, hotels, motels, KOAs, amusement parks, FAX machines, pagers, long-distance telephone service, printing services, business forms, business magazines, tapes, prescription drugs, and plenty more. Call *PD News* for more details, or see Appendix C for information about NASE.

PD News also offers the following services, some of which are available to nonsubscribers:

- Résumé Typing Service—*PD News* will type your résumé, send you a proof copy for your approval, and then send you the final copy. $20 per page. (Nonsubscribers can use this service for $75 per page).
- Weekly Résumé Mailing Service—*PD News* mails résumés each week to the current week's advertisers—about 150 companies. You can include your résumé for $40 for the first page, $5 for additional pages.
- Monthly Résumé Mailing Service—*PD News* will mail a résumé to all the advertisers in the *Directory of Technical Service Firms*—about 800 companies—for $195 for the first page and $20 for additional pages.
- Hot List of Contract Personnel—Add your name, address, telephone number,

and job category to a list that *PD News* mails to about 150 agencies. $3 for one week, or $10 for four weeks. Nonsubscribers can also use this service. Free to PD Passport members.
- Computer Bulletin Board—A 24-hour computer bulletin board with the latest job listings. Also includes an open forum on which you can leave messages, a computerized Hot List service, and a list of all the agencies with listings in the latest issue of *PD News*. You can also write "letters to the editor" and notify *PD News* of your address change. Free to subscribers and PD Passport members. At the time of this writing, *PD News* was also planning to add a directory of technical service firms to its bulletin board.
- Instant Shopper Retrieval System—This service is free to anyone, even non-

subscribers. A contractor can send his name and address to *PD News* to be added to its database. If an agency is looking for a particular person—perhaps someone who has worked for them in the past—they can call *PD News* with the name and *PD News* searches the database for the current address. If you would like to be added to the database, call *PD News* at (512) 250-8127.

Contract Employment Weekly (C.E. Weekly)

C.E. Publications
P.O. Box 97000
Kirkland, WA 98083-9700
(206) 823-2222
FAX (206) 821-0942

A weekly publication with display ads and listings of contracts throughout the nation. 80 pages, about 500 agencies per week. $45 for 52 issues, $25 for 15 issues, $10 for 5 issues, $3 for one issue. The $45 and $25 subscriptions include the annual *Directory of Contract Service Firms*, listing about 1,000 agencies. Nonsubscribers can purchase the directory for $10.

Other services:

- *Highway to Success* Video—a 23-minute video explaining the contract employment field, $40. Available to nonsubscribers.

(The following services are available only to subscribers.)

- Résumé Mailing Service (Weekly)—Send a copy of your résumé and *C.E. Weekly* will copy it and mail it to about 500 agencies the following Friday. $120 for a one-page résumé, $20 for additional pages.
- Résumé Mailing Service (Monthly)—*C.E. Weekly* will copy and mail your résumé to about 1,000 agencies at the end of the month. $240 for a one-page résumé, $40 for additional pages.
- Résumé Mailing Service (Combined)—*C.E. Weekly* will copy and mail your résumé to all the agencies in their Directory and all the advertisers in the most recent *C.E. Weekly*, about 1,050 agencies. $260 for a one-page résumé, $50 for additional pages.
- Résumé Mailing Service (Regional)—Select the region you want to mail your résumé to (West, South Central, North Central, etc.). The agencies are from the annual *Directory of Contract Service Firms*. One-page résumé per region, $50, additional pages $10.
- Résumé Typing Service—You provide the information, and *C.E. Weekly* will type and print a résumé. $25 per page. You will receive the original typed résumé for this price, but if you want a PMT copy (suitable for reproducing on a bond-type copier), include an extra $8.
- Résumé Printing Service—You supply the résumé, or use the Résumé Typing Service. *C.E. Weekly* will print copies for the following single-page rates: 50 copies, $13; 100 copies, $17. Additional pages are $8 for 100. Prices include cost of mailing résumés to you.
- Tele-Résumé Service—*C.E. Weekly* will FAX your résumé to any company you want. $2 per page. Send *C.E. Weekly* a résumé and a $10 deposit. Then, when you need to FAX a résumé, just phone them and give them the FAX number.
- Mailing Labels—*C.E. Weekly* sells mailing labels. *C.E. Weekly* Advertisers (from the most recent issue), $12 for about 500 labels. Directory Advertisers, $24 for almost 1,000 labels. Directory Advertisers (Regional), $7 per region, or $3 for all the foreign-address labels.
- Hot Sheet of Contract Personnel—Add your name, address, telephone number,

and job category to a list that *C.E. Weekly* mails to about 1,050 agencies. $5 for one week. Free to unemployed subscribers, or those who expect to be out of work within two weeks (only one free listing per month).
- Mini-Ads in the Hot Sheet—2 large-type lines. Free to unemployed subscribers, or those who expect to be out of work within two weeks.
- Display Ads in the Hot Sheet—Quarter-page, $50; half-page, $100; full page, $200.
- Address Change in the Hot Sheet—List your new address and phone number in the Hot Sheet. $2.
- Advertise in *C.E. Weekly*—"The Weekly Log" (subscribers seeking work), near the front of the publication: five lines, one time, $15; three times, $13 per week; twelve or more times, $10 per week. Eighth-page Display Ad: one time, $140; three times, $105 per week; twelve or more times, $70 per week.
- Computer Database—A 24-hour computer bulletin board with the latest job listings. Also includes an electronic mail system. Free to subscribers.
- Business Cards—With *C.E.* logo. $29 to $46 for 1,000, depending on the ink colors.
- Résumé Acknowledgment Cards—Include these with mailings to agencies. The agencies can check the boxes on the card and drop it in the mail. Printed with your name, address, and phone number. $18 for 100, additional 100s for $10.
- Employment Information Sheets—Checklists to use when you talk with agencies. $3.50 for 25 sheets.
- Mail Forwarding Service—A permanent mail forwarding address, for while you are "on the road."

Consultants' and Contractors' Newsletter, and Job Express

Consultants' and Contractors' Publications
105 North Main Street
Boonton, NJ 07005
(800) 836-0667 (national)
(800) 648-9926 (NJ)
(201) 299-1535
(FAX) (201) 402-1823

If you live in the New York-New Jersey-Connecticut-Pennsylvania area, you should take a look at these publications. *Job Express* is a twelve-page newsletter that is published twenty-two times a year. It lists contracting and consulting jobs, and carries subscribers' ads. Many of the listings—probably most—are from agencies. I called a technical writer who had run an ad; he told me that it had "brought a number of responses" (from agencies) and he once found work through the ad. In fact, he had run the ad twice and was planning to do it again. The ads are $25 for one listing of *unlimited length*. Four listings—they don't have to be consecutive—are $70. *Job Express* also carries such extras as market surveys of rates and information about local job fairs and exhibitions.

The Consultants' and Contractors' Newsletter is published monthly. It's twelve pages long, half of which is filled with business-card-size ads from agencies and services. The rest contains short articles on consulting—interviews with people in the business, local news, contracting-law updates, book reviews, and so on—and a few ads.

The same company also publishes *The Consultant's Ye**ow Pages*, a directory with about 1,000 listings. The entries are firms and government agencies that use data processing personnel, and include a contact name and telephone number.

A one-year subscription to both *Job Express* and *The Consultants' and Contractors' Newsletter* is $66. Six months is $39.50. You also get *The Consultant's Ye**ow Pages* free with the one-year subscription. If you want to buy the *Ye**ow Pages* without subscribing, it's $22.95 plus $1.95 for postage for the first copy, but only $0.75 for the second—I suppose they figure it will be photocopied anyway! You can also buy *CCN's Guide to Consulting and Contracting* ($19.25 plus $1.95) and *The Freelancer's Guide to Taxation* ($26.95 plus $1.95). (The first is 34 pages and a bit weak; I haven't seen the second.)

Job Express and the newsletter focus on the Eastern Seaboard, but have a few listings in other areas. How would I use these publications if I lived in that area? I would probably advertise in *Job Express*, and use the listings and ads to build my own contact lists. (I wouldn't just wait until I saw an ad for a technical writer's job, because they probably won't appear that often—most agencies deal mainly with programmers, it seems, and only occasionally with writers).

Appendix B
Technical Service Firms

This appendix lists the names and telephone numbers of a few technical writing companies and large technical service firms. As you know from the rest of the book, I recommend that you build your own list of agencies in the area in which you want to work, but you can use this list to make initial contacts and get a feel for the market. Call some of the agencies and talk to them for a few minutes. You can discuss the demand for your skills and may get an idea of how much agencies are paying and where the work is. And, who knows, you may end up with a contract without even trying.

I selected most of these agencies because they are large national agencies with work throughout the United States, even the world in some cases. In all cases the address given is for the corporate headquarters, but when you call you should ask for the address of their local offices; if you mail résumés, many of these companies will automatically forward your résumé to the office nearest to you. Many of the companies have 800 numbers, so you can do a little research without paying for the phone bill.

The first two firms are small companies, but of special interest because they work almost exclusively with technical writers.

SkuppSearch, Inc.
Holly Skupp
580 Sylvan Avenue
Englewood Cliffs,
  NJ 07632
(201) 894-1824
FAX (201) 894-1120
(permanent placement of
  technical writers)

Techwrights, Inc.
540 Route 10 West,
  Suite 314
Randolph, NJ 07869
(201) 786-7244
(a company that places
  consultants/contractors
  in the Boston to
  Washington area)

Aide, Inc.
POB 6746
Greenville, SC 29606
(803) 244-6123

B & M Associates
Corporate Office
199 Cambridge Road
Woburn, MA 01801
(617) 938-9120
(800) 225-8710

Belcan Technical Services
Corporate Office
POB 429138
Cincinnati, OH 45242
(513) 489-4300
(800) 543-4543

Butler Service Group
110 Summit Avenue
Montvale, NJ 07645
(201) 573-8000
(800) 526-0320

CDI Corporation,
  Corporate
  Headquarters
Ten Penn Center
Philadelphia,
  PA 19103-1670
(215) 569-2200
(800) JOB-LINE
  (562-5463)

Consultants and
  Designers, Inc.
360 West 31st Street
New York, NY 10001
(212) 563-8400

Ewing Technical Design
Royalpar Industries
40 South Street
POB 10479
West Hartford, CT 06110
(203) 249-6311

The Franklin Company
600 Reed Road
Broomall, PA 19008
(215) 356-1010
(800) 523-4948

General Devices, Inc.
207 East Main Street
POB 667
Norristown, PA 19404
(215) 272-4477

Gonzer Associates, L.J.
1225 Raymond Boulevard
Newark, NJ 07102-2919
(201) 624-5600
(800) 631-4218
(Eastern states only)

International Technical
   Services
141 Central Avenue
POB 239
Farmingdale, NY 11735
(516) 694-4433
(800) JOB-TIME
   (562-8463)

Kirk Mayer, Inc.
11801 Mississippi Avenue
Los Angeles, CA 90025
(213) 272-6131

Lehigh Design
14120 McCormick Drive
Tampa, FL 33626
(813) 855-9411

MiniSystems
124 West Figueroa Street
Santa Barbara, CA 93101
(805) 963-9660
(800) 445-5506
   (outside CA)

Nelson, Coulson and
   Associates, Inc.
333 West Hampden, #507
Englewood, CO 80110
(303) 761-7680

Nesco Design Group
14120 McCormick Drive
Tampa, FL 33626
(813) 855-9411

Oxford and Associates,
   Inc.
75 Pearl Street
Reading, MA 01867
(617) 944-6200
(800) 426-9196

PDS Technical Services
POB 619820
Dallas, TX 75261
(214) 621-8080
(800) 777-9372

Peak Technical Services,
   Inc.
3424 William Penn
   Highway
Pittsburgh, PA 15235
(412) 825-3900
(800) 284-2841

Pollack and Skan, Inc.
120 West Center Court
Schaumberg, IL 60195
(312) 359-4949
(800) 544-7817
   (outside IL)

Quantum Resources
POB 35630
Richmond, VA
   23235-0630
(804) 320-4800

Ray Rashkin Associates,
   Inc.
1930 South Alma School
   Road #D107
Mesa, AZ 85210
(602) 897-2479
(800) 543-6076

Salem Technical Services
Attn: National Recruiting
1333 Butterfield Road
Downers Grove, IL 60515
(312) 990-8800
(800) 323-7200

S&W Technical Services, Inc.
8122 Datapoint Drive #930
San Antonio, TX 78229
(512) 699-1080

SEI Technical Services
7725 Little Avenue
Charlotte, NC 28226
(704) 542-7100
(800) 331-1618

Superior Design
250 International Drive
POB 9057
Williamsville, NY 14231-9057
(716) 631-8310

TAD Technical Services Corp.
639 Massachusetts Avenue
Cambridge, MA 02139
(617) 868-1650
(800) 225-5776 (outside MA)
(800) 842-1417 (MA only)

Tech/Aid
POB 128
Needham Heights, MA 02194
(617) 449-6632
(800) 225-8956

TRS, Inc.
POB 26147
Greenville, SC 29616
(803) 297-3110
(800) 522-5627

Volt Technical Services
101 Park Avenue
New York, NY 10178
(212) 309-0300
(800) 367-8658

Western Technical Services
301 Lennon Lane
Walnut Creek, CA 94598-9280
(415) 930-5300

Yoh Company, H.L.
1818 Market Street
Philadelphia, PA 19103
(215) 299-8400
(800) 523-0786, xtn 8400

Appendix C
Associations

The most important association for technical writers is

The Society for Technical Communication
901 North Stuart Street
Arlington, VA 22203
(703) 522-4114
(703) 522-2075

The STC has 15,000 members—15 to 20 percent of whom are independents—in 125 chapters (mostly in North America). The organization is "devoted to the advancement and the theory and practice of technical communication," but it's also a great way to meet technical writers in your area. Most chapters have monthly meetings, and many have annual seminars and competitions and publish local newsletters. The national organization itself publishes a quarterly journal (*Technical Communication*), a newsletter that appears ten times a year (*Intercom*), and a quarterly newsletter for student members (*Interchange*). It also publishes a number of books and manuals—*Basic Technical Writing* for $15 ($25 for nonmembers), for example, and the *Guide for Preparing Software User Documentation* for $10 ($15 for nonmembers). See Appendix E for more information.

The STC also holds an annual conference, awards scholarships, and offers research grants. You can also buy insurance at a discount through Mutual of Omaha if you are an STC member: "Forty to 50 percent off disability insurance," M of O told me, and 5 percent off the medical plan.

The STC can even help you find work. Just meeting your colleagues is an important way to find job leads, of course, but many local chapters also maintain a "job bank" that lists local jobs and contracts. The STC publications often carry ads for jobs, and the STC's computer bulletin board system includes a nationwide list of jobs. You can also purchase the national membership directory (some chapters publish a local directory), and there is a Consulting and Independent Contracting Professional Interest Committee with 600 members. It publishes a quarterly newsletter, *The Independent Perspective*, has a Consulting Forum on the STC's bulletin board, conducts regional conferences, and is planning to publish "kits" such as a "useful contracts" kit and one about setting rates. The STC even has a forum on CompuServe, the national computer information network. (Anyone who uses CompuServe can join this forum. See "Working from Home Forum," Section 10, Independent Writers.)

The 1992 fees for the STC are $85 plus a one-time $10 enrollment fee, and $30 for students (no enrollment fee).

You might also check out The National Association for the Self-Employed:

The National Association of the Self-Employed (NASE)
2328 Gravel Road
Fort Worth, TX 76118-6950
(800) 232-6273

This is a 200,000-member, strong association with a wide range of services including health and disability insurance. Call for information. (As this book was going to print, I learned of serious complaints about NASE's health policy. See *Home Office Computing*, Jan. 1992.)

Appendix C

You could also track down other local organizations; for example, if you work in telecommunications, you may want to join a society related to that industry. These associations can be an excellent source of leads. Go to your library and ask if they have a directory of local associations. You can also check the following books:

The Encyclopedia of Associations, Gale Research Company
The National Trade and Professional Associations of the United States, Columbia Books
Consultants and Consulting Organizations Directory, Gale Research Company

If you still can't find an organization related to your profession, maybe you should start one. It may even be possible to start one as a profit-making venture; see Howard Shenson's book *The Complete Guide to Consulting Success* for a discussion of starting a professional association.

Here are a few other associations that may be of interest.

American Association of
  Computer Professionals
72 Valley Hill Road
Stockbridge, GA 30281
(404) 474-7874

American Association of
  Professional Consultants
9410 Ward Parkway
Kansas City, MO 64114
(816) 444-3500

American Consulting
  Engineers Council
1015 15th Street NW
Washington, DC 20005
(202) 347-7474

Association for
  Computing Machinery
11 West 42nd Street
New York, NY 10036
(212) 869-7440

Association of Computer
  Professionals
230 Park Avenue
New York, NY 10169
(212) 599-3019

Association of Data
  Communications Users
POB 20163
Bloomington, MN 55420
(612) 881-6803

Association of Electronic
  Cottagers
POB 604
Sierra Madre, CA 91024
(818) 355-0800

Association of
  Information Systems
  Professionals
1015 North York Road
Willow Grove, PA 19090
(215) 657-6300

Fastbreak Syndicate, Inc.
POB 1626
Orem, UT 84059
(801) 785-1300
(writers, illustrators,
  artists, etc.)

Freelance Network
POB 36838
Miracle Mile Station
Los Angeles, CA 90036
(213) 655-4476
(technical writers,
  photographers,
  illustrators, artists, etc.)

Independent Computer
  Consultants' Association
933 Gardenview Office
  Parkway
St. Louis, MO 63141
(314) 997-4633

Professional and Technical
  Consultants Association
1330 South Bascom
  Avenue, Suite D
San Jose, CA 95128
(408) 287-8703

Appendix D
Contractor's Checklist

Use the first part of this checklist before you begin looking for work. Make copies of the "Negotiating with the Agencies" and "On the Road" sections of the checklist and use them to take notes when you talk to the agencies.

Before Looking for Work

Have you calculated how much you earn per hour?

$_____ per hour

(Remember that this is an estimate only. How much you "earn" is a relative number, depending on the cost of replacing your benefits; see the explanation in chapter 8.)

Have you selected a medical insurance policy?
Have you selected a long-term disability policy?
Have you selected a life insurance policy?

Beginning Your Search

USING AGENCIES

Have you made a list of all the agencies?
        Check: Colleagues
                     Yellow Pages
                     Contractors' Publications
                     Classified Ads
Have you written your cover letter and résumé?
Have you printed your mailing labels?

NEGOTIATING WITH THE AGENCIES

Before accepting a contract you must know the answers to these questions; remember to ask both the agency and company, where applicable (for example, dress code or amount of overtime):

Is the job a permanent position or a contract?
Is the agency looking for contractors or employees?
What industry is the job in?
What type of work?
How long is the contract?
How much is the agency paying?
Where is the contract?
Who is the client?
How much overtime is available?
How much overtime is expected?
Is overtime paid at time-and-a-half?
Does the agency have a medical insurance policy?
        How much is the policy? (single, married, children)
        How much is the deductible?
        How much is the out-of-pocket?

What percentage does it pay after the deductible?
> (Eighty percent is normal.)

Does it include dental?

Does it include vision?

Does the agency have a long-term disability policy?
- How much is it?
- Is it included with the medical?
- How much coverage does it provide?
- What is the waiting period?

Does the agency pay for vacations or sick leave?
- How long do you have to work before getting paid leave?
- What is the ratio of work days to free days?
  - (e.g., work 130 days, get six days off)
- Are paid-leave days ever vested? E.g., if you work for nine months will they pay all free days or only those earned in the first six months?

Does the agency pay a mileage/travel allowance?
- How much?

Does the agency have a 401(k) pension plan?
- How long must you be employed to be eligible?

Personal:
- Dress code?
- Smoking allowed, banned?
- Company cafeteria?

_____

_____

_____

On the Road

Does the job pay a per diem?
- How much per week?

Is there a state/local income tax?
- How much?

Is there a state/local sales tax?
- How much?

How much more will general living expenses be?

How much will accommodation cost?

Will the agency pay any moving expenses?

ANYTHING ELSE

(Add your own special requirements here.)

Contractor's Checklist

Looking for Independent Contracts

Have you called your "key" contacts?

_____

_____

_____

_____

_____

Have you called all the members of your network? Friends, recent colleagues, etc.
Have you called old employers?
Have you called old colleagues?
Have you called local professional-society job banks?
Have you checked the newspaper classified ads?
Have you used computer bulletin boards?
Have you managed to find a list of your peers?
    Have you "cold-called" this list?
Have you been to any local job fairs?
Have you found a list of companies that can use your services?
    Have you "cold-called" this list?

Appendix E
Bibliography

Books About Writing, Technical Writing, and the Writing Business

Cash Copy. Dr. Jeffrey Lant (JLA Publications, 50 Follen Street, Suite 507, Cambridge, MA 02138, (617) 547-6372). *Don't waste money on advertising or direct mail until you have read this book from the self-promotion guru. Buy directly from him and you'll never get off his mailing list, but you might find some other interesting books.*

Clear Technical Writing. John A. Brogan (McGraw-Hill). *Excellent; you have to get this one.*

The Complete Guide to Writing Software User Manuals. Brad M. McGehee (Writer's Digest Books). *A good primer on writing computer user guides.*

The Elements of Style. William Strunk, Jr., and E.B. White (Macmillan Publishing Co.). *Every list of writing books has to include this one, doesn't it?*

How to Write a Computer Manual. Jonathan Price (Benjamin/Cummings Publishing Co.). *Covers writing, testing, and revising your work, and how to recommend software changes. Very good.*

How to Write a Usable User Manual. Edmond H. Weiss (ISI Press). *Stodgy, but a good overview of some of the tech-pubs production. Some companies use this book as their tech-writing bible.*

Secrets of a Freelance Writer: How to Make $85,000 a Year. Robert W. Bly (Dodd, Mead & Company). *Lots of good ideas about selling your services. And yes, you can make (or exceed) $85,000 a year, but it will take hard work.*

The Technical Writer's Handbook. Matt Young (University Science Books). *A humorous dictionary of terms and misuse.*

Technical Writing: A Reader-Centered Approach. Paul V. Anderson (Harcourt Brace Jovanovich). *A big book that seems to cover everything from grammar to page layout.*

Technical Writing for Business and Industry. Patricia Williams and Pamela Beason (Scott, Foresman, & Co.).

Writing Effective Software Documentation. Patricia Williams and Pamela Beason (Scott, Foresman, & Co.). *Lots of examples of page layout.*

Society for Technical Communication publications. These are published by the STC (see Appendix C for their address). *I haven't used any of these, so I can't recommend them, but I'm sure some of them could be useful. I have included both the member's price and the nonmember's price.*
    Basic Technical Writing
    Proposals and Their Preparation

Technical Communication and Ethics
Technical Editing: Principles and Practices
Guide for Beginning Technical Editors
How to Teach Technical Editing
Slaying the English Jargon
Teaching Technical Writing
Technical and Business Communication
Guide for Preparing Software User Documentation
The Scientific Report: A Guide for Authors
Freelancing and Consulting Information Kit
Technical Translation Information Kit
Freelance Nonfiction Articles
How to Develop a Format for Any Publication
How to Function as a Schizoid Editor
How to Train Your Boss in Technical Communication

OTHER MATERIAL

Complete Guide to Consulting Success. Howard Shenson (Enterprise Publishing, Inc., 725 North Market Street, Wilmington, DE 19801). *An excellent overview of consulting. This book has a one-year, "no questions asked" guarantee.*

The Complete Guide to Health Insurance. Kathleen Hogue (Walker & Co.).

The Consultant's Kit: Establishing and Operating Your Successful Consulting Business. Dr. Jeffrey L. Lant (JLA Publications, 50 Follen Street, Suite 507, Cambridge, MA 02138, (617) 547-6372).

Consulting: The Complete Guide to a Profitable Career. Robert E. Kelley (Charles Scribner's Sons).

Freelancing—The First 30 Days. Bill Coan (Coan & Company, 606 Kessler Drive, Neenah, WI 54956). *An excellent guide to the "nuts and bolts" of finding contract work. This book has a money-back guarantee.*

Going Freelance: A Guide for Professionals. Robert Laurance (John Wiley & Sons).

Health Insurance Made Easy—Finally. Sharon Stark (S.L. Stark, 7525 Norwood, Prairie Village, KS 66208, (913) 383-9039). *Self-published. Ms. Stark used to be a benefit approver and customer service representative for a medical insurance company. This is a detailed description of all the medical insurance benefits and exclusions, and what to look for when buying a policy.*

How Do I Pay For My Long Term Health Care? Maya Altman (Berkeley Planning Association). *This is aimed at retirees and pre-retirees.*

J. K. Lasser's Your 19xx Income Tax. The J. K. Lasser Institute (Simon & Schuster). *Reissued towards the end of each year. This is a good guide to your taxes. Includes a special year-end supplement, mailed to your home, providing last-minute tax updates and a telephone hot line with taped information on over fifty subjects.*

The New Freelancer's Handbook: Successful Self-Employment. Marietta Whittlesley (Simon & Schuster). *Although written for freelancers working in the arts (writers, actors, dancers, etc.), this is an interesting book covering everything from coping with job-related stress to how to get onto a TV game show.*

The Only Other Investment Guide You'll Ever Need. Andrew Tobias (Bantam). *The sequel to* Still the Only Investment Guide You'll Ever Need. *A good way to make sure you use your newfound wealth wisely.*

Payment Refused. William Shernoff (Richardson & Steirman). *If you are interested in some of the dirty tricks that insurance companies use to avoid paying your claims, read this. It gives a little information on protecting your rights, also.*

The Psychology of Call Reluctance: How to Overcome the Fear of Self-Promotion. George Dudley and Shannon Goodson (Behavioral Science Research Press). *Teaches you to overcome any reluctance to sell yourself; especially useful if you really hate cold-calling.*

Shopping For Health Care in Confusing Times. Henry Berman & Louise Rose (Consumer Reports).

Small-Time Operator. Bernard Kamoroff (Bell Springs). *This bestseller is a clear, concise guide to keeping your accounts and paying taxes.*

Still the Only Investment Guide You'll Ever Need. Andrew Tobias (Bantam). *A good way to figure out what to do with all the extra money you are going to earn, and a useful discussion about life insurance. You may also want to read* The Only Other Investment Guide You'll Ever Need *(above).*

What's Wrong With Your Life Insurance? Norman F. Dacey (Macmillan Publishing Co.).

Winning the Insurance Game. Ralph Nader and Wesley Smith (Knightsbridge Publishing). *Misnamed; it should have been* Trying to Win the Insurance Game. *The way the insurance industry has the odds stacked against you, there's not much chance of actually winning. You're going to need all the help you can get. This one's a good one.*

Appendix F
Co-ops, Correspondence Courses, Courts, Insurance, and the IRS

Co-op America (from p. 91—insurance programs)
2100 M Street, NW
Washington, DC 20063
(202) 872-5307
(800) 424-2667

Correspondence Course (from p. 28—technical writing/communications)
University of California Extension
University of California
Berkeley, CA 94720
(415) 642-8245

Internal Revenue Service (from p. 125—for tax bulletins)
(800) 424-3676

SelectQuote Insurance Services (from p. 94—insurance programs)
140 Second Street
San Francisco, CA 94105
(800) 343-1985

Small Claims Court Citizens Legal Manual
(from p. 47)
HALT, Inc.
201 Massachusetts Avenue, NE, Suite 319
Washington, DC 20002

Workers Trust (from p. 91—insurance programs)
POB 11618
Eugene, OR 97440
(503) 683-8176
(800) 447-2345

Index

A
Accommodations, prices for, 72
Active voice, 30–31
Advantages of freelancing, 39–43
Agencies. *See* Technical service agencies
American Almanac of Jobs and Salaries, 22
Amortization, 124
Associations, 147–148
Attitude, 99

B
Background of technical writers, 22–26
Bad-contract cycle, avoidance of, 61–62
Benefits
   buying, 44, 89–96
   as promises, 40
   replacement costs for, 52–58
Books, on technical writing and business, 29, 152–154
Boring work, 17
Bulletin boards, 113
Business capital
   building, 105–106
   financial responsibility and, 44
Business cards, 110
Business contacts, 108, 110–111
Business lunches, 110
Business travel expenses, tax deductions for, 118–119

C
Card file, on companies and contacts, 107, 108
Catastrophic care policy, 93
Change, handling, 51
Client(s). *See also* Companies
   fair rates and, 73–74
   interview with, 80–82
   prospective, talking with, 112
   rate of, 114–115
   relationship with contractor, 97–98
   taxes and, 129
COBRA (Consolidated Omnibus Budget Reconciliation Act), 54, 90–91
Colleagues
   cold calling, 113
   common feeling among, 13
   as help for finding agency, 62–63
College courses, listings of, 28
College students, as technical writers, 21
Command table, 32
Commuting, IRS deductions for, 117–118
Companies. *See also* Client(s)
   calling to get next contract, 112
   with contract, 69
   finding businesses that can use your service, 113–114
   hiring independent contractors through an agency, 78–79
   hiring independent contractors without an agency, 79
   reasons for avoiding agencies, 75–76
   that only use agencies, 115–116
Company office, work habits/actions for, 98–99
Competition, among agencies, 61, 77
CompuServe, 113
Computer-book market, 16–17
Computer books, writing, 136–137
Concise writing, 30
Confidence, 60–61
Consolidated Omnibus Budget Reconciliation Act (COBRA), 54, 90–91
Consultant(s)
   becoming, 130, 133–134
   as contractor, 33–34
   definition of, 134
   fees for, 134–135
   independent contracts and, 35
   vs. contractor, 98
Consultants' and Contractors' Newsletter, 142–143
Consulting work, finding, 135–137
Contacts, business, 108, 110–111
Contract(s)
   in agency negotiations, 67
   independent, 84–88
   legal aspects of, 129–130
   length of, 68
   location of, 69
   next, steps in getting, 112–114
   restrictive covenant of, 83–84
   signing, 72–73
   with technical service agency, 35
   types of, 87
   writing, consideration factors for, 87–88
Contract Employment Weekly (C.E. Weekly), 66, 141–142
Contractor(s). *See also* Freelancer(s)
   agency negotiations and, 67–68
   checklist before you look for work, 149–151
   good, not working with agencies, 77
   income of, 15
   physical appearance of, 99
   rehiring, 78
   relationship with client, 97–98
   types of, 35
   vs. consultant, 33–34
   why companies use, 36
Contractors' publications, 63, 139–143. *See also specific publications*
Co-ops, 155
Correspondence courses, 155
Credit unions, 96
Cross-references, 31

Index

D
Day care, 56
Dental insurance, 54
Depreciation, 124
Diagrams, 31
Direction, 45
Direct mail, 135–136
Disability insurance, 93–94
Diversity
  in background of technical writers, 22–26
  of technical writing jobs, 14, 20–21
Documents, by technical writers, 14

E
Earnings. *See* Income
Editors, professional, 32
Education, 55
Employee discounts, 55–56
Employees
  exempt, 101
  salaried, of technical service agencies, 35
  of technical service agency, 35
  temporary, 33, 35
  vs. contractors, 67–68
Employees nonexempt, 101
Employer contributions, to pensions and tax-free savings plans, 55
Employer of contractor. *See* Client(s); Companies
English graduates, as technical writers, 21
Exempt employees, overtime rates and, 101
Exemptions, employee, 102–104
Ex-military personnel, as technical writers, 21
Experience, 36–37, 41

F
Fair Labor Standards Act, 79, 102–103
Feature or command table, 32
Fees. *See* Rate of payment
Female technical writers, 15–16
FICA, 56–57, 123
Finding work. *See* Job hunting
Fixed fee, 47
Fixed-fee contract, 87
Flexi-time schedule, 130
401(k) pension plan, 71
Freelance business, building, 105
Freelancer(s). *See also specific types of freelancers*
  characteristics of, 48–51
  getting feel for market, 49–50
  incompetent, 97
  independent, 35. *See also* Independent contractor(s)
  as salesperson, 45, 49, 99–100
Freelancing
  definition of, 33
  disadvantages of, 44–47
  endorsement of, 43
  as stepping-stone, 41–42
  in technical professions, 38
  Three-Step Method of, 9–11, 37
  types of, 34–36
"Front" corporation, 130

G
Go-It-Alone Plan, 11
Gossip, 49–50

H
Headings, 31
Health club, 55
Health Maintenance Organization (HMO), 92
Helping people, building networks and, 110
Home offices, 117, 120
Hourly rates
  for freelancers, 14–15
  for hours worked, 40
  of permanent job, when to calculate, 53
  for technical writers, 34, 50
Hours worked per year, 57
Human relations skills, 51

I
Income. *See also* Money
  annual, 39
  calculation of earnings from permanent job, 52–58
  of freelancers, 14–16
Incorporation, 126, 130
Independent contract(s)
  checklist for, 151
  description of, 84–87
  fee basis, 35
  hourly rate, 34
Independent contractor(s). *See also* Sole proprietor
  hiring through agencies, 78–79
  hiring without agency, 79–80
  IRS criteria for, 127–128
  profits of, 36
  taxes and, 123–124
  types of, 34–35
Independent Retirement Account (IRA), 55
Indexes, 32
Industry, type of, 68
Initiative, 99
Insecurity, feelings of, 45–46
Insurance, 155
Interviews, 80–82, 109
IRA (Independent Retirement Account), 55
IRS, 129, 155

J
Jargon, 30
Job banks, 109, 113
Job description, 14
Job express, 142–143
Job fairs, 109
Job hunting
  for consultant, 135–136
  for freelancer, 41
  spending time on, 46
  steps for finding next contract, 112–115
  for technical writer, 26–27
Job interviews, 109
Job offers, 60
Job satisfaction, 38
Job security, 41, 42
Job-shopper, 33, 35
Job shops, 34. *See also* Technical service agencies
Joint employment, 78–79, 131
Journalists, as technical writers, 21

K
Keogh plan, 55, 95–96
Knowledge, of contract market, 106

Index

L
Language, clarity of, 30
Learning
   about contract market, 106
   quickly, 51
Leaving a freelance job, 41
Legal aspects, 129–130
Length of contract, 68
Life, balanced view of, 40–41
Local tax, 72
Location of contract, 69
Long-term disability insurance, 53, 54, 70
Long-term working relationships, 44
Looking for work. *See* Job hunting

M
Mailing costs, 66
Mailing labels, 64
Mailings, preparation of, 64–65
Male technical writers, 15
Medical insurance
   affordability of, 91–92
   of agency, 70
   buying, 89–93
   cost to replace, 53–54
   group types, 91–92
   preexisting condition restrictions, 90
   of spouse, 91
Mileage, tax deductions for, 117–118
Mileage allowance, 71
Military veterans, medical insurance for, 90
Money. *See also* Income
   building business capital, 105–106
   discussing with captive employees, 100
   handling abilities, 48
   as job advantage, 39
Moving expenses
   payment of, by agency, 72–73
   tax deductions for, 122–123

N
National Association of the Self-Employed (NASE), 147
Negotiations, with agencies, 149–150
Networking
   building a network, 108–111
   definition of, 107
   fee setting and, 135
   gossip and, 49–50
   information needed from, 107–108
Newspaper classified ads, 63–64, 109, 113
Non-Employee Compensation tax form (Form 1099NEC), 129
Nonexempt employees, overtime rates for, 101
Nonwords, 30

O
Office politics, 42–43, 45
On-the-job training, 28
Organizational skills, poor, 18
Overtime rates
   availability of overtime and, 69–70
   employee exemptions and, 102–104
   as time-and-a-half, 101–102

P
Parties, 110
Payment(s)
   defaulting on, 47
   methods for, 47
   rate of. *See* Rate of payment
PD News, 66, 139–141
Pension plans, 45, 55, 71, 95–96. *See also specific pension plans*
Per diems
   availability of, 71–72
   tax deductions for, 118–119
Permanent employment, 67, 78
Personal considerations, in negotiations with agency, 71
Personnel recruiters, 67, 110
Phone calls, to get next contract, 112
Physical appearance, 99
Problems in technical writing, 16–17
Profession, 20
Professional associations, 108, 147–148
Profit, of agency, 75
Proofreading, 32
Pseudo-salary, 73, 102–104
Public Law, 101–583, 104

Q
Qualifications, 78
Quality of work, 50

R
Rate of payment. *See also* Salaries
   estimating, 134–135
   estimating charge for project, 134–135
   fair, 73–74
   for good contractors, 77
   hourly. *See* Hourly rates
   negotiation with agency, 68–69
   for other freelancers, learning about, 107–108
   for overtime. *See* Overtime rates
Readers of technical manuals, 29–30
Reading level, 31
Recruiters, personnel, 67, 110
Recurrent disability clause, 94
Redundancies, 30
Reference aids, 32
References, checking, by agency, 76
Referrals, 42
Rehiring of contractor, 78
Reputation, 51, 75, 135
Respect, lack of, 17
Restrictive covenants, in contracts, 83–84
Résumés
   emphasis in, 22–23
   keeping on file with agency, 61
   misrepresentation in, 18
   preparation for mailing, 64–65
Revenue Act of 1978, safe harbor clause, 129
Revenue Ruling 87-41, 127
Risk, 9

S
Salaries. *See also* Rate of payment
   agency negotiations and, 69
   of female technical writers, 16
   overtime and, 102, 103
   for technical writers, 14–16

Index

Salesperson, freelancer as, 45
Savings, 44
Savings plans, tax-free, 55, 94–96. *See also specific plans*
Schedule C, 124
Search, using agencies, 149
Secretaries, as technical writers, 21
SelectQuote, 94, 155
Self-esteem, 41
Self-motivation, 50
Selling yourself, 49
SEP (Simplified Employee Pension plan), 55, 95
Sick leave, 52, 70–71
Simplified Employee Pension plan (SEP), 55, 95
Skills, 27
SkuppSearch, Inc., 27
Small Claims Court, 47, 155
Social Security tax, 56–57, 123
Society for Technical Communication (STC)
   description and address of, 147
   job banks, 109
   joining, 26
   pointers from, 29–32
   survey of technical writers, 14, 20
Sole proprietor, 35, 123–124. *See also* Independent contractor(s)
Spell-checkers, 32
Spouse, medical insurance of, 91
Starting your business, 20–27
State tax, 72
STC. *See* Society for Technical Communication
Subchapter S corporations, 130
Subheadings, 31
Supplemental employment, 35

T
Table of contents, 31–32
Tables, 31
Tax(es)
   business travel expenses and, 118–119
   doing yourself, 116
   filing, 120
   freelance employees and, 120–122
   home office deductions and, 120
   independent freelancer and, 123–124
   mileage deductions and, 117–118
   moving expenses deductions and, 122–123
   paying promptly and correctly, 131
   per diems and, 118–119
Tax books, 125
Tax forms, 125
   Form 4562, 124
   Form 1099NEC, 129
Tax-free savings plans, 55, 94–96. *See also specific plans*
Tax-homes, IRS criteria for, 119
Tax preparers, 116
Tax rates, 72
Tax Reform Act of 1986, 128
Tax Schedule C, 124
Technical personnel, as technical writers, 21–22
Technical service agencies
   companies that only use agencies, 115–116
   competition among, 61, 77
   contacting, reasons for, 60–62
   contracts with, 35
   description of, 34
   employee of, 35
   finding, 62–64
   freelance employees of, taxes and, 120–122
   helping people in, 110
   IRS and, 128
   large vs. small, 67
   listing of firms, 144–146
   long-term disability insurance of, 70
   mailings to, 64–66
   medical insurance of, 70
   negotiations with, 67–74
   placement of writers, 60
   providing leads for, 42
   rates of, 15, 77
   relationships with, 108
   salaried employee of, 35
   share of company payment, 77–78
   specialties of, 60
   starting your own, 130
   technical writing problems and, 18
   unethical, 75–79
   working with, 66
Technical writers, other. *See* Colleagues
Technical writing classes, 26–27
Temporary employees, 33, 35
Term life insurance, 54–55, 94
Three-Step Method, 9–11, 37
Time, spending on job hunting, 46
Time factors, length of contract, 68
Time off, 40
Training
   educating yourself, 46
   on-the-job, 28
Transfers, professional, 47
Travel
   business, tax deductions for, 118–119
   as career advantage, 41
   checklist, 150
   life on the road, 66
   questions to ask when working out of town, 71–73
Traveling expenses, 72–73
Type of work, 68

U
Uncertainty, feelings of, 45–46, 48–49
Unemployment pay, 47

V
Vacation pay, 52, 70–71
Vacations, 46
Variety, as job advantage, 39
Vision insurance, 54

W
Work environment, 98–99
Workers compensation insurance, 47
Writers, problems with, 17–19

Y
Yellow page categories, for finding agency, 63